The Effectiveness of Mathematics Teaching in Primary Schoo'

AF755893

The Effectiveness of Mathematics Teaching in Primary Schools: Lessons from England and China provides a unique insight into the mathematics classrooms of these two countries and arrives at a time when the world is eager to know how Chinese learners consistently excel at learning mathematics and other core subjects.

Showcasing the kinds of teaching methods that work within and across countries, this book presents a rich collection of views, including those from teachers, their native colleagues, their foreign colleagues and the researcher, regarding the quality of mathematics teaching today. Interweaving scientific results about teaching and learning evaluations with multiple perspectives of various roles in and out of the classroom, Miao and Reynolds offer insights into how and why different approaches of teaching have led to different learning outcomes in mathematics internationally. Building on rigid and robust analyses of the most up-to-date data in England and China, the book indicates that it is through changing teaching rather than changing teachers that mathematics learning can be improved, because it is what teachers do in the classroom that really makes a big difference.

Containing four decades of wisdom from the field of teaching effectiveness research, this book is essential reading for all who want to improve the quality of mathematics teaching worldwide. This book is particularly relevant for educational researchers, postgraduate students and teachers, as well as school leaders, policymakers and parents.

Zhenzhen Miao is Lecturer at the College of Elementary Education, Jiangxi Normal University, China. She completed her PhD under the supervision of Professor David Reynolds at the University of Southampton and had worked as a research assistant at the University of Aberdeen in the fourth year of her PhD. Zhenzhen's research interests are inspired by her experiences of being a mother and a former award-winning mathematics and English teacher in China. She has a strong interest in the STEM subjects, especially mathematics.

David Reynolds (CBE) is Professor of Education at Swansea University. He has published 20 books and more than 200 articles and book chapters. He is one of the founders of the field of School Effectiveness and School Improvement (SESI), the SESI Journal and the International Congress for School Effectiveness and Improvement (ICSEI). David has advised, lectured and/or consulted in over 50 countries over the years.

Routledge Research in International and Comparative Education

For a full list of titles in this series, please visit www.routledge.com

This is a series that offers a global platform to engage scholars in continuous academic debate on key challenges and the latest thinking on issues in the fast growing field of International and Comparative Education.

Books in the series include:

Citizenship Education and Migrant Youth in China
Pathways to the Urban Underclass
Miao Li

International Service Learning
Engaging Host Communities
Edited by Marianne A. Larsen

Educational Borrowing in China
Looking West or Looking East?
Charlene Tan

Nationalism and History Education
Curricula and Textbooks in the United States and France
Rachel D. Hutchins

Global Literacy in Local Learning Contexts
Connecting Home and School
Mary Faith Mount-Cors

Dialogue in Places of Learning
Youth Amplified in South Africa
Adam Cooper

Faculty Development in Developing Countries
Improving Teaching Quality in Higher Education
Edited by Cristine Smith and Katherine E. Hudson

The Effectiveness of Mathematics Teaching in Primary Schools

Lessons from England and China

Zhenzhen Miao and David Reynolds

Routledge
Taylor & Francis Group
LONDON AND NEW YORK

First published 2018 by Routledge

2 Park Square, Milton Park, Abingdon, Oxfordshire OX14 4RN
52 Vanderbilt Avenue, New York, NY 10017

Routledge is an imprint of the Taylor & Francis Group, an informa business

First issued in paperback 2018

Copyright © 2018 Zhenzhen Miao and David Reynolds

The right of Zhenzhen Miao and David Reynolds to be identified as
authors of this work has been asserted in accordance with sections 77 and 78
of the Copyright, Designs and Patents Act 1988.

All rights reserved. No part of this book may be reprinted or reproduced or utilised in any form or by any electronic, mechanical, or other means, now known or hereafter invented, including photocopying and recording, or in any information storage or retrieval system, without permission in writing from the publishers.

Notice:
Product or corporate names may be trademarks or registered trademarks, and are used only for identification and explanation without intent to infringe.

British Library Cataloguing-in-Publication Data
A catalogue record for this book is available from the British Library

Library of Congress Cataloging-in-Publication Data
A catalog record for this book has been requested

ISBN: 978-1-138-68032-6 (hbk)
ISBN: 978-0-367-17863-5 (pbk)

Typeset in Bembo
by Apex CoVantage, LLC

We dedicate the book to Every Child in this world past, present and future for they are the ultimate teachers who teach teachers how to teach them best.

All grown-ups were once children . . . but
only few of them remember it.
– Antoine de Saint-Exupéry, *The Little Prince*

Contents

Figures

Tables

Acronyms and abbreviations

CHC	Confucian-heritage-cultural
CN	China or Chinese
CN-FG	Chinese focus group
CPD	continuing professional development
DfE	Department for Education (UK)
DfEE	Department for Education and Employment (UK)
EER	educational effectiveness research
EMT	effectiveness of mathematics teaching
EN	England or English
EN-FG	English focus group
ETN	the Effective Teachers of Numeracy project
ETS	the Educational Testing Service
FIAC	the Flanders Interaction Analysis Categories
FIMS	the First International Mathematics Study
GDP	gross domestic product
GVA	gross value-added
IAEP	the International Assessment of Educational Progress
ICCAMS	the Increasing Competence and Confidence in Algebra and Multiplicative Structures project
IEA	the International Association for the Evaluation of Educational Achievement
ISERP	the International School Effectiveness Research Project
ISTOF	the International System for Teacher Observation and Feedback
itvQ	interview question
IWB	interactive whiteboard
LPS	the Learner's Perspective Study
M1	Method 1
M2	Method 2
M3	Method 3
M4	Method 4
M5	Method 5
M6	Method 6
MEPP	the Mathematics Enhancement Project Primary

MET	the Measure of Effective Teaching project
NAEP	the National Assessment of Educational Progress
OECD	the Organisation for Economic Cooperation and Development
Ofsted	the Office for Standards in Education
ORACLE	Observational Research and Classroom Learning Evaluation
OTL	opportunity to learn
PACE	the Primary, Assessment, Curriculum and Experience project
PC	personal computer
PCK	pedagogical content knowledge
PE	Physical Education
PISA	the Programme of International Student Assessment
PQ	pupil questionnaire
PRC	People's Republic of China
QUAL	qualitative
QUAN	quantitative
RQ1	Research question 1
RQ2	Research question 2
RQ3	Research question 3
SATs	Standard Assessment Tests
SD	standard deviation
SE	school effectiveness
SEN	special educational needs
SER	school effectiveness research
SES	socio-economic status
SI	school improvement
SIMS	the Second International Mathematics Study
SMK	subject matter knowledge
SPSS	Statistical Package for the Social Sciences
STEM	science, technology, engineering and mathematics
TA	teacher assistant
TALIS	the Teaching and Learning International Survey
TER	teacher effectiveness research
TIMSS	the Third International Mathematics and Science Study
TIMSS	Trends in International Mathematics and Science Study
TQ	the teacher questionnaire
UK	the United Kingdom
US	the United States of America
WB	whiteboard

1 The effectiveness of mathematics teaching study

Chapter overview

- Rationale and questions
- Contexts and schools
- What's in the book?

In this increasingly digitalised and globalised world, the quality of education in mathematics and other STEM subjects predicts the competitiveness of a country's future workforce. International comparative studies on achievement have repeatedly identified the disparity in mathematics performance between East Asia and the West (Mullis et al., 1997; Mullis, Martin, & Foy, 2008; Mullis, Martin, Foy, & Arora, 2012; Mullis, Martin, Gonzalez, & Chrostowski, 2004; Reynolds & Farrell, 1996) and between China and England (Lapointe, Mead, & Askew, 1992; Lapointe, Mead, & Phillips, 1989; OECD, 2010, 2013).

Evidence from empirical studies has consistently proved that schools make a difference and teachers have a much larger impact upon pupils' achievement than do schools (Hill & Rowe, 1995; Reynolds & Teddlie, 2000; Willms, 2000). It has been argued that it was necessary to get into the classroom and see what actually happened if one wanted to improve learning (Hiebert et al., 2003). This study thus set out to (1) investigate teaching factors that work across England and China and make a difference in pupils' mathematics learning outcomes, (2) understand how different roles in and beyond classrooms view mathematics teaching across two countries and (3) interpret the interconnections between hard evaluations of teaching and learning and soft perspectives of practitioners and the researchers.

This introductory chapter reflects upon the importance of mathematics, explains the research rationale and questions of the study, puts you in the national and local contexts where the teachers and pupils in the study come from and finally offers an overview of the remaining chapters of the book.

Rationale and questions

First of all, why *mathematics*? As one of the most important bodies of knowledge, mathematics has developed across cultures and over millennia. It is the oldest school of human thought, alongside philosophy (Krantz, 2010). The mathematical historian, Carl Boyer, affirmed in his work, "that the beginnings of mathematics are older than the oldest civilisations is clear" (Boyer, 1968, p. 7). As far as archaeologists have discovered, the formal application of mathematics can be traced back to the era of Old Egypt and Old Babylonia around 3200 BC (Nuffield Foundation, 1994, Ch. 2). Since the era when Arabians drew together the two lines of mathematics – the East (ancient India) and the West (ancient Greece) – through translation and introduced it to Europe, mathematics has gradually become an international body of knowledge (Boyer, 1968; Hodgkin, 2005).

Throughout history, mathematics has come from humans' observation of the world and guided human beings in tackling real-world problems, providing powerful tools to the development of all sciences. In today's world, mathematics remains a crucial part of almost every branch of science, including physics, chemistry, biology, architecture and electronic and computer science. For its importance in the past, present and future, mathematics is regarded, in schools across the world, as a core subject that provides essential foundations for other school subjects. The importance of mathematics makes it important that children be well educated in the subject. Such an importance also results in the increase of the number of educational studies focusing on the quality or process of mathematics education over the past 50 years. The following section will explain why a new study like this is so much needed.

International comparative surveys, conducted by the International Association for the Evaluation of Educational Achievement (IEA) and the Organisation for Economic Cooperation and Development (OECD), have been regularly evaluating educational outcomes of core subjects, such as mathematics and science, cross-nationally. Although the target ages of pupils those international surveys choose to focus on are different from one another, they all include mathematics as an assessed subject. The results of international assessments draw much attention and raise questions as to how educational effectiveness differs amongst nations and cultures (Hiebert & Stigler, 2004). The number of international surveys on maths learning outcomes is much greater than that on maths teaching practices which have been mainly looked at on a within-country basis.

Empirical evidence from teacher effectiveness research (TER) consistently reveals positive correlations of certain teacher behaviours with pupil learning outcomes/gains in mathematics and other core subjects (Muijs & Reynolds, 2011). However, over the past four decades, such evidence has been limited to the national level. In this increasingly globalised world, there is a lack of international attempts to reshape educational research agendas, evaluate which teaching variables "travel" across nations and which do not, "tap the full range of variation", generate "more complex, sensitive and multi-layered explanations" of differences in learning outcomes, and subsequently inform classroom innovations, as for the case of school effectiveness research (SER) (see Reynolds, Stringfield, Teddlie, & Creemers, 2002b, pp. 5–6).

This study, therefore, seeks to fill the research gap by collecting and analysing data across countries to find up-to-date evidence on the effectiveness of mathematics teaching (EMT) that could travel confidently across countries as mathematics itself did, does and will undoubtedly continue to do.

Applying a combination of convenience and stratified sampling strategies, the study collects data in two countries – England and China – and involves maths teachers and their pupils aged 9 to 10 (i.e.Year 5 in England and Grade 4 in China). In each country, a city from the medium-to-high level stratum was chosen, and within each city (Southampton, England and Nanjing, China), schools were sampled from the average stratum. Whilst detailed sampling techniques, procedures and rationales are illustrated in Chapter 4, the next section of this chapter draws a picture of the research contexts.

As will be reviewed in Chapter 3, the area of TER is bounded by national borders, has a strong quantitative characteristic, but lacks rich descriptions (Reynolds, Sammons, De Fraine, Townsend, & Van Damme, 2011; Reynolds et al., 2002b). To make meaningful contribution to the area of TER and to children's education in mathematics worldwide, this study is intended to fulfil the research purposes as follows:

- To evaluate and correlate the quantity and quality of mathematics teaching with learning outcomes cross-nationally;
- To collect multiple voices on the effectiveness of mathematics teaching;
- To connect *hard* measurement with *soft* views and throw light on the improvement of mathematics teaching and learning within and across countries.

These purposes then lead to the formulation of research questions to be followed and the design of research methods in Chapter 4.

Motivated by the learning gaps of school mathematics between pupils across countries in international performance surveys, the study aims at measuring the teacher effects in two countries' mathematics lessons on pupil learning outcomes with internationally validated instruments and collecting multiple views regarding the effectiveness of mathematics teaching. A list of methods was applied to collect and analyse data so as to answer the following questions:

RQ1 What are the correlations between various teaching approaches and learning outcomes in mathematics across two countries?

RQ1a What is the quantity and quality of mathematics teaching in classrooms across two countries?

RQ1b How do children perform in two standardised mathematics tests?

RQ1c How does teaching correlate with learning across mathematics classrooms in the two countries?

RQ2 How do different roles in and beyond classrooms perceive the effectiveness of mathematics teaching in the two countries?

RQ2a How does the researcher interpret the quality of mathematics teaching in the two countries?

RQ2b How do individual teachers view the effectiveness of mathematics teaching in general and that of the specific lesson they delivered for observation?

RQ2c How do groups of teachers view the effectiveness of two mathematics lessons – one from England and the other China?

RQ3 How does the rigid evaluation of teaching and learning (RQ1) connect with the diverse perceptions of different roles about the effectiveness of mathematics teaching (RQ2)?

This book is a report of the EMT project through which we seek answers for the above questions.

Contexts and schools

Now we will have a close look at the research contexts – the participating nations, cities and schools and the comparability between them.

England and China

As follows, a brief introduction is given to England and China, in terms of geography, population, culture, politics, education and economy. Comparability of the two countries is explained as well.

Geography

England is the largest of the four countries in the United Kingdom of Great Britain and Northern Ireland (the UK or Britain). The area of England is 130, 281 km^2 – more than a half of the total area of the UK (Office for National Statistics, 2013b). It is bordered by Scotland to the north and Wales to the west and surrounded by the Irish Sea to the northwest, the North Sea to the east, and the English Channel to the south. Divided by the 33-km English Channel from France, England is the nearest country of the UK to mainland Europe. The People's Republic of China (generally called China) is located in East Asia, between latitudes 18° and 54°N, and longitudes 73° and 135° E. With Beijing as its capital, China has a total area of 9.6 million km^2, which makes it the world's second-largest country by land and the third or fourth by total area. It shares land borders with 14 countries and maritime borders with eight countries (PRC Gov., 2013).

Demography

The latest 2011 census sees a population of 53,012,456 living in England. With about a half of the UK's land, England is accommodating 83% of its total population (63,182,178) according to the 2011 census (Office for National Statistics, 2013a). In China, there are 56 ethnic groups – 55 minorities (Zhuang, Uyghur, Hui, Manchu, Miao, Tibetan etc.) and one ethnic majority (Han). In 2010, China had a population of 1,339,724,852, with 16.6% of them being aged 0–14, 70.1% 15–59 and 13.3% 60 and above (National Bureau of Statistics of China, 2011).

Culture

English culture is not much different from British culture as a whole. It has the characteristics of an island country, the features of a Christian-based society, and traces of a former world superpower. It has its own unique taste in literature, music, architecture, folklore, philosophy etc. All of these have various connections with Christian-oriented beliefs and the priority given to freedom and democracy. England nurtured world-famous writers, such as Geoffrey Chaucer and William Shakespeare, and singer-songwriters, such as John Lennon and Paul McCartney (Keet, 2012). Starting its evolution along two old rivers, the Yangtze River and the Yellow River, Chinese culture has its roots settled in Confucianism, Taoism and Buddhism. The three streams of philosophy, particularly Confucianism, have formulated the cultural phenomena that Chinese predominantly value harmony, family responsibility, education, good morality and diligence, yet historically generally lacking critical thinking and the courage to challenge authorities. After 1949, traditional culture in mainland China was altered by politics in many ways. The post-1979 era has seen a gradual decrease of such political influence over its traditional culture.

Politics

With London as its capital, England is a non-devolved constituent country within the constitutional monarchy and parliamentary system of the UK. England is directly governed by the British Parliament – one of world's oldest parliaments. China is a centralised socialist country which is run by one party, the Communist Party of China. The electoral system is hierarchical, allowing limited democratic processes within the party and at the local village level.

Education

The Department for Education (DfE) in England is the national department responsible for education of people from basic to higher education. England provides compulsory education for people from the age of 4 up to 16. Schools are inspected by the Office for Standards in Education (Ofsted) and graded into four levels: outstanding, good, satisfactory and inadequate. English primary schools employ generalist teachers who teach almost all main subjects. Children often are grouped by ability within and/or across classes. There are often teacher assistants (TAs) working alongside class teachers to provide extra support to those with special education needs (SEN). During a school day, children often have to travel between classrooms for different subjects' learning according to their ability levels. The Ministry of Education is in charge of education at all stages in China including higher education. China provides nine years' compulsory education to children (aged 6 to 15) from primary through lower secondary phases. Teaching Research Offices (教研室) at district (county) and municipal levels are responsible for school inspections, focusing particularly on the process and performance of teaching and learning. Nonetheless, performance details of primary schools are generally not open to the public, partially to prevent parents from selecting to live in certain catchment areas on purpose. Chinese primary schools employ specialist teachers and take the mixed-ability approach within and across classrooms in all subjects. Upon entrance pupils are randomly allocated to fixed classes where different subject teachers come and deliver lessons. There is no such role as a TA in Chinese schools – every teacher ought to teach independently.

Table 1.1 shows a comparison between the two countries' schooling phases along the timeline of children's ages. Those cells shaded in light grey show the stages of compulsory education in each country. The bold age group, i.e. 9- to 10-year-olds at the primary phase, is the target age group of the EMT study. Pupils aged 9–10 years in China are attending primary Grade 4, whereas pupils at the same age are taught in Year 5 in England.

Economy and comparability

The first industrialised country on the planet, today the UK is still one of the world's largest economies. Its average GDP per capita ($38,514) is ranked the

Table 1.1 Overview of the educational phases in two countries

China		*Age*	*England*	
	Lower Grade	3–4		Preschool
Kindergarten	*Middle Grade*	4–5	*Reception*	
	Upper Grade	5–6	*Year 1*	
	Grade 1	6–7	*Year 2*	
	Grade 2	7–8	*Year 3*	Primary
Primary	*Grade 3*	8–9	*Year 4*	
	Grade 4	**9–10**	**Year 5**	
	Grade 5	10–11	*Year 6*	
	Grade 6	11–12	*Year 7*	
	Grade 7	12–13	*Year 8*	
Lower Secondary	*Grade 8*	13–14	*Year 9*	Secondary
	Grade 9	14–15	*Year 10*	
Upper Secondary/ Professional Institutions	*Grade 10*	15–16	*Year 11*	
	Grade 11	16–17	*Year 12*	Sixth Form/Further Education
	Grade 12	17–18	*Year 13*	
Higher Education/Professional Institutions		18+	Higher Education/Further Education	

21st in the world (The World Bank, 2013). China is now the world's second-largest economy after the US. Nonetheless, its GDP per capita ($6,076 in 2012) is ranked 86th out of 182 countries by the International Monetary Fund (Wikipedia, 2013). Despite the disparity of GDP per capita between the two countries, it is reasonable to compare the relatively developed regions in China with the equivalent areas in England.

Southampton and Nanjing

General information

Sitting on the south coast of England, Southampton has a total area of 51.81 km^2 (Southampton City Council, 2013) and a population of 236,900 (Hampshire County Council, 2013). The proportion of children aged 0–14 is 16.4%, that of people aged 15–59 66%, and that of people being 60 or above 17.6%. The ratio between White British and other ethnic groups is 77.7% to 22.3% (*ibid.*). With a glorious maritime history, today Southampton is still home to luxury cruise ships and the fourth largest British port by tonnage. Other pillar industries include car and aircraft manufacture, petrochemicals, electronic engineering etc. There are 79 state schools, four independent schools and three independent special schools in Southampton. Of all state schools, 21 are infant schools (for children aged 4–7), 16 junior schools (7–11), and 24 primary schools (4–11), 12 secondary schools (12–16) and six special schools (DfE, 2013). Like the case of other regions in England, primary school admissions give priorities to those who have siblings in the same school, live close to the school, are in special care, have faith in a specific religion or went to a feeder school.

Nanjing is located at the lower reaches of the Yangtze River. Once an ancient capital of the country and now the capital of Jiangsu province, it has a total area of 6,598 km^2. In 2010, Nanjing had a population of 8.005 million, with 6.47 million living in the urban area. The ethnic group, Han, is 98.56% of the city's population, the rest of which are from 50 minority groups. Nanjing has been a prosperous city since ancient times. There are 344 primary (for children aged 6–12), 160 lower secondary (12–15) and 95 upper secondary (15–18) schools in Nanjing (Nanjing Education Bureau, 2011). To ensure equal distributions of educational resources, it is a national policy that every child should be enrolled to the primary and lower secondary schools (i.e. the compulsory phase) of their catchment areas. Schools at the compulsory stage are not allowed to select prospects in any forms. In Nanjing, the Education Bureau does not publish or rank publicly the performance of any primary schools. The purpose is to prevent parents from rushing to buy properties in catchment areas of 'best' schools.

Economy

Southampton was ranked the 27th out of 45 cities in the UK in terms of competitiveness in 2013 (Huggins & Thompson, 2013). Its GVA (£52,800 per capita in 2013) was ranked 16th out of 63 cities (Centre for Cities, 2015). Overall, Southampton is representative of the medium-to-high group of cities in England socio-economically. Nanjing's GDP per capita was ranked 11th out of 36 major cities (in population terms) in China (National Bureau of Statistics of China, 2013). Its pillar industries include electronics, petrochemical, steel and automobile (China Knowledge, 2014). Nanjing is representative of the medium-to-high level cities in China in socio-economic terms.

Comparability

The national and municipal data and information shows the international comparability of Southampton and Nanjing, in socio-economic terms, with each of them representing the medium-to-high level cities in its home country. There are, of course, differences in terms of culture, politics, geography and demography, which is unavoidable in international comparisons but will be taken into consideration in the interpretation of the research findings later.

The participating English schools

The English sample comes from three primary schools, EN-A, EN-B and EN-C. All Year-5 teachers along with their pupils (excluding several who opted out) from each school took part in the study.

Two of the schools were founded in the 1950s and one in the 1990s. All three schools are junior schools which provide education to children aged 7 to 11 in Years 4 through 6 (i.e. Key Stage 2). All schools were graded as 'Good' according to their latest Ofsted reports up until 2012. In almost every classroom, there is a generalist teacher assisted by a TA (occasionally more than one). School EN-B is unique in that Teachers EN3 and EN4 do not have a TA whereas Teacher EN5 has two to three TAs in her low-ability class.

School EN-A has about 170 pupils organised in 2 classes of each year group. Both teachers (EN1 and EN2) in Year 5 have taken part in the study. The school has ability-based grouping within maths classes but does not have maths sets across classes. School EN-B has approximately 290 pupils and three classes per year group. Children are set in maths both within and between classes. All three teachers (EN3, EN4 and EN5) from Year 5 have taken part in the study. School EN-C has about 520 pupils organised in five classes in each of the four years. Like School EN-B, the school employs ability grouping in maths within and between classrooms. All Year-5 teachers (EN6, EN7, EN8, EN9 and EN10) have participated in the study.

A typical English school starts at 9 am and children have two periods (approx. 1 hour 40 minutes per period with a 15-minute break in between) in the morning and one period in the afternoon. A period usually consists of two lessons about two different subjects without a break. For example, the teacher might start an English lesson immediately after a maths lesson. Lunch usually starts at around 12 pm and lasts about 50 minutes. The formal school time ends at around 3:15 pm. Some schools have after school clubs every weekday, whereas others only on certain days, for example, Fridays. The class teacher is responsible for all subjects' teaching except the foreign language, usually French. Some may also teach after school clubs. The classroom is normally also the teacher's office – typically a corner (about the size of 2–3 m^2) beside the IWB with a desk, a PC, some drawers, and other essential facilities.

The participating Chinese schools

The Chinese sample includes all Grade-4 maths teachers and their pupils from four primary schools in Nanjing. Schools CN-A and CN-B, both founded in 2007, are relatively newer than the other two Chinese schools. School CN-A has six classes in each of six grades (years) and approximately 1,200 pupils and about 80 members of teaching staff. School CN-B has six classes per grade (year) with about 1,400 pupils and 110 teachers. School CN-C was founded in 1934 and rebuilt on the same site in 2006. With a team of about 60 teachers, it caters for about 650 pupils from Grades one through six. Due to urban development and commercial expansion, the surrounding community is shrinking. Thus, the school has two classes in Grade 4. Founded in 1994, School CN-D has four classes in each grade and caters for slightly over 700 pupils with a team of 55 teachers. As a matter of policy, all schools take the mixed-ability approach.

Conventionally, a maths teacher – like teachers of other core subjects – teaches two parallel classes in the same grade (same content, same pace). In this study, the Chinese team includes three teachers (CN1, CN2 and CN3) from School CN-A, three (CN4, CN5 and CN6) from CN-B, one (CN7) from CN-C, and two (CN8 and CN9) from CN-D.

A typical primary school day in Nanjing starts at 8 am and ends at 3 pm. There are four sessions in the morning and two in the afternoon. Each session lasts 35 to 40 minutes. There is a 10-minute break between sessions 1 and 2, 3 and 4, 5 and 6 and a 25-minute break between sessions 2 and 3 for broadcast gymnastics and, in some schools, also eye exercises. Lunch starts at around 12 pm, and session 5 starts at about 1:30 pm. There are usually after school clubs offering non-academic activities on campus. Teachers teaching the same subject share an office where a desk, a PC and other essential facilities are provided for every teacher. Together, they form the teaching research group (教研组) of the specific subject. In many schools, teachers of the same subject are often further grouped by grade. For example, all Grade-4 maths teachers may share an office.

What's in the book?

This book has nine chapters. The paragraphs below provide an overview of the remaining nine chapters to come:

After this introductory chapter, Chapter 2 offers an insight into pedagogical theories and practices in school mathematics and a bird's-eye view of international studies on mathematics learning and teaching. We will then move on to the evolution of educational effectiveness research and its subfield – teaching effectiveness research – in Chapter 3.

In Chapter 3, we review sequentially the evolution of educational effectiveness research (EER), early American literature on teaching effectiveness research (TER), British literature on TER and other nations' literature on TER. The limitations of prior studies on mathematics pedagogy and international comparisons of mathematics learning and teaching (as reviewed in Chapter 2) and TER among countries consolidate the research purposes and help shape the research questions.

Chapter 4 follows the line of research purposes and questions, establishes the integrated paradigm and sets within it multiple methods for collecting and analysing data across classrooms and nations. The theoretical framework of the project is formulated to guide the subsequent data collection and analysis processes. We then detail the data-collection-and-analysis methods and procedures and the way we integrated findings across methods.

Chapter 5 presents the background information about the children participating in the study and gives an overview about teachers and their professional selves. All information was gathered through pupil and teacher questionnaires.

In Chapter 6, with two internationally validated observation systems, lessons were evaluated and analysed systematically. One observation system focuses on the quantity of teaching, and the other the quality of teaching. The next part tells the overall results and detailed patterns of children's learning outcomes as measured with a paper of 40 maths items derived from the released TIMSS 2003 item bank for fourth graders (9- to 10-year-olds). Part four correlates teaching (both quantity and quality) with learning (two maths tests) and shows what works whereas what doesn't.

A wide range of rich voices has arisen from unstructured observations, video-stimulated interviews with teachers and video-stimulated focus groups with teachers. In Chapter 7, we will hear these diverse voices from different roles in and beyond classrooms in and across the two countries about the effectiveness of mathematics teaching.

Chapter 8 addresses the research questions, brings together results and findings of the study and interweaves them at different layers. This is where research questions, research methods and research findings join each other in the integrated paradigm in representing and informing the reality of maths teaching and learning.

Chapter 9 points out the contributions, indications and limitations of the study and foresees potential directions for future research and teaching innovations. With four decades of research evidence and the updated evidence from the EMT study, this chapter provides reliable ways for improving maths teaching across nations.

2 Growing apart

Worldwide pursuit of effective maths education

Chapter overview

- Mathematics pedagogy: theories and practices
- Facts and factors in international maths learning gaps
- International comparisons of maths learning outcomes
- International comparisons of maths teaching practices

Much effort has been devoted worldwide to the effectiveness of education and in particular mathematics education over the past century. In this chapter, we will look at the effort and outputs generated by two communities of researchers focusing on: (1) mathematics pedagogy and (2) international comparisons of mathematics learning and teaching.

Mathematics pedagogy: theories and practices

Pedagogy is defined as 'the method and practice of teaching' (Oxford Dictionaries, n.d.). In other words, it is about *how* to teach. Studies on pedagogy therefore share the same vision – seeking the *best* way of teaching – with studies in teaching effectiveness research (TER) which will be reviewed in Chapter 3. Works on mathematics pedagogy so far are generally either theories or empirical studies driven by theories. In this section, we will mainly trace the roots of dominating pedagogical theories in school mathematics and reflect upon some of the most investigated pedagogical aspects.

Progressive teaching

Progressive education has its root in trendy thoughts and theories around the beginning of the 20th century – most specifically, John Dewey's thoughts on democracy and education (Dewey, 1916) and Jean Piaget's theory of cognitive development (Piaget, 1950; Piaget & Inhelder, 1969). Founder of pragmatism, Dewey emphasised the role of freedom in the classroom where children should be actively experiencing and acquiring knowledge through hands-on activities rather than passively accepting the knowledge transferred by the somewhat authoritative teacher. Piaget believed that children were born with an existing ability to learn, that their learning – a combined result of children's physical growth and their experience – went through a fixed sequence of stages, and that therefore the classroom should be child-centred and individualised with minimal teaching input, so that every individual child's developmental needs could be met.

Deweyan and Piagetian theories had transformed the landscape of school education in the West in the 20th century and formed the heart and soul of progressive education which now retains its influence in Western schools. The fundamental change it brings to education is that the role of teachers in the classroom is to ensure that individual children are free to explore the intended knowledge independently with minimal formal teaching input. The underlying assumption of progressive teaching is that in this way children can progress through stages of learning with maximal freedom and minimal external intervention. Alternative names for progressive teaching include constructivist, discovery, problem-based, experiential and inquiry-based teaching (Kirschner, Sweller, & Clark, 2006).

In a progressive setting, teachers are not encouraged to deliver the knowledge directly, and children are expected to acquire the knowledge themselves through exploration at their own pace. Progressive teaching is normally time-consuming and results are not ideal. Research has shown either a loss of learning or a low level of knowledge acquisition amongst children taught in this way, although the low ability learners tend to enjoy the explorative learning process (Clark, 1982, 1989).

The UK has been at the forefront of the implementation of progressive education. In the late 1960s, the Plowden Report team drew the blueprint of the country's educational reform by referring to Piaget's theory (Central Advisory Council for Education, 1967). Since then, progressive education, as opposed to traditional education, has become the norm in British schools. The philosophy behind it even dominated the landscape of research literature on learning and teaching at the time. As Bennett argued, there was little on teaching but a growing body of literature on learning and "a few fairly crude theories of teaching, based not on the observation of practice, but loosely on American conceptions of democracy" (Bennett, 1987, p. 67).

In fact, at the primary school stage, explorative learning appears to work against the way the human brain processes information over the short and long terms (Atkinson & Shiffrin, 1968). Learners can only store a small amount of information over a short period and their learning of new knowledge has to be built on prior knowledge that is stored in their long-term memory. In "unguided environments" (Kirschner et al., 2006, p. 80), the amount of information stored in long-term memory sharply distinguishes novice learners from expert learners in that the latter have rich experience and sufficient knowledge in their long-term memory which helps reduce their cognitive load and assists them to make sense of challenging and complex learning content, whereas the former encounter enormous difficulties in the absence of both internal and external guidance. Therefore, at the early stage of learning, external guidance – such as learning worked examples – is necessary, as Kirschner et al. put it:

> It emphasizes the importance of providing novices in an area with extensive guidance because they do not have sufficient knowledge in long-term memory to prevent unproductive problem-solving search.
>
> (Kirschner et al., 2006, p. 80)

The naïve intention of progressive teaching in school mathematics, physics or chemistry is to create an environment for novices to obtain knowledge in a way a mathematician, physician or chemist generates knowledge in their career lives. However, the premise for a professional mathematician or scientist to work in an explorative and *productive* way is that they have already accumulated a solid base of knowledge (internal guidance) in their fields, whilst this is not the case for a child (extremely talented excluded) in the classroom. "The practice of a profession is not the same as learning to practice the profession" (Kirschner et al., 2006, p. 83). In mathematics, teachers cannot largely rely on explorative and self-guided activities, since every bit of the body of mathematics knowledge was probably the result of one mathematician's or a number of mathematicians' lifelong exploration. Children have only one chance to learn a *curriculum* of school mathematics over a dozen of years or so.

Democracy and freedom should be advocated, but it cannot be assumed to be a magic bullet for the teaching of specific school subjects, such as mathematics. There has to be a scientific way of teaching mathematics which must be evidence-based. Since children have only one chance, teachers and educators should not apply untested theories to them. Unfortunately, this happened and is still happening. Progressive education has been experimented in the West for quite a long time in the 20th century, parallel to which is the decline of pupils' understanding of fundamental knowledge in mathematics (e.g. see ICCAMS Maths, 2015).

Teaching and ability grouping

As a way of differentiating teaching, streaming or setting has also existed for a long time in schools in Western countries, such as the UK and the US. In England, streaming became the norm in the 1950s, and schools generally streamed pupils into different ability classes and taught them all subjects in a fixed class. Despite the reversing trend favouring the mixed-ability approach around the 1980s, ability grouping has retained its dominance in English schools. National policies have every now and then reinforced this climate. Notably, the year 1997 saw the publication of the White Paper, *Excellence in Schools*, which declared that setting was "worth considering in primary schools" and "should be the norm in secondary schools" (DfEE, 1997, p. 38). Since then, setting has been widespread up and down the country. In English schools, there are now different forms of setting that group pupils by ability either within classes, or across classes, or both within and across classes.

Widely practised, setting is, however, generally found negative towards children's self-esteem. Collecting data in six British schools, Boaler, William, and Brown (2000) found not only the negative influences of setting in maths on the emotion of pupils in both high-ability and low-ability sets (i.e. classes) but also the pedagogical challenge that maths teachers encountered as they strove to differentiate teaching. On the contrary, Kulik and Kulik's (1992) meta-analysis suggested positive effects of grouping by ability upon academic performance and denied the negative effects that setting might pose both academically and emotionally. Whilst the effect of setting on learning remains slightly debatable, ability itself does not seem to be the single major predictor in deciding which set a child might be assigned to, since the SEN and social background of a child also plays a considerable role in set assignment (Muijs & Dunne, 2010).

Teacher knowledge

Around the late 1970s, inspired by the Piagetian theory, researchers in maths education had sought to identify pupils' levels of mathematical understanding in specific areas, such as decimals, ratio, algebra (Hart et al., 1981). Formative assessment was considered helpful for teachers to understand pupils' mathematical thinking and to improve their own subject matter knowledge.

About teacher knowledge, Shulman's work is one of the few that educational researchers in maths and many other school subjects are very likely to recall first. In his seminal work, Shulman (1986) fiercely criticised the then standards for teacher evaluation and teaching and school effectiveness research in the US for their specific interest in teaching process and outcomes and their ignorance of the *content* which he categorised as (1) subject matter knowledge (SMK), (2) pedagogical content knowledge (PCK) and (3) curricular knowledge. Amongst the three domains of teacher knowledge, PCK was regarded as the most important, since the grasp of the SMK, for example mathematics, did not in itself mean the capacity of making others understand and grasp the knowledge equally well. Four more types of teacher knowledge were later added: (4) general pedagogical knowledge, (5) knowledge of learners, (6) knowledge of educational contexts and (7) knowledge of educational purposes (Shulman, 1987).

In both research and practice in maths education, teachers' subject-related knowledge has been widely emphasised and explored. Through task-based interviews, Ma (1999) compared American and Chinese maths teachers' understanding of basic subject matter and pedagogical content knowledge and found that Chinese teachers not only showed a better understanding of problem-solving procedures (usually also with several alternative solutions) but also generated richer and deeper accounts of the underlying concepts and better ways of teaching them than American teachers.

In the US, empirical evidence suggests that teachers' mathematical content knowledge poses significant impact on pupils' mathematics performance

(Hill, Rowan, & Ball, 2005). In the UK, studies have shown the British trainee teachers' lack of SMK and PCK in teaching mathematical topics, such as subtraction, gradient and geometry (Jones, Mooney, & Harries, 2002; Rowland, 2010). Because of the lack of competence in mathematics and therefore the dearth of confidence in the subject themselves, many English primary teachers are not confident in *teaching* the subject. A case study suggests that teachers may become more confident after taking part in continuing professional development (CPD) programmes that aim at improving their subject matter knowledge (Hodgen & Askew, 2007). Another study (Howe et al., 2015) found higher learning gains in the classes of teachers who had participated in CPD programmes focusing on PCK in rational numbers. The study claimed that the effect appeared to be greater over a longer term. Nonetheless, there was a lack of systematic analyses of the whole teaching process.

Overall, research efforts so far are still more related to SMK than PCK, and research findings available seem fuzzy and cannot thoroughly address the questions Shulman posed three decades ago:

> How does the successful college student transform his or her experience in the subject matter into a form that high school students can comprehend? . . . How does the teacher prepare to teach something never previously learned? How does learning *for* teaching occur?
>
> (Shulman, 1986, p. 8)

Despite the insufficient output over the past few decades, research on maths PCK has broadened its vision to throw the spotlight on both teachers' understanding (internal) and their teaching processes and outcomes (external). This is commendable for at least two reasons: (1) the teaching process is the very means by which teachers' PCK is put into action; (2) the outcome of teaching and learning is the touchstone of teachers' capacity in communicating the subject content to pupils in a comprehensible manner.

Teaching with coherence

Teaching with coherence is another element that has been repeatedly discussed in work on maths pedagogy over the recent years (Ma, 1999; Schmidt et al., 1996; Shimizu, 2007; Stigler & Hiebert, 1999). A coherent lesson is expected to build close connections between segments of a lesson and between a series of lessons. Such connectedness should exist at the conceptual level rather than merely the surface level. Thus, strong subject matter and pedagogical content knowledge seem to be crucial for a teacher to teach coherently. Coherence is one of the things that maths teachers pay particular attention to in East Asian countries, such as China (Chen & Li, 2010) and Japan (Sekiguchi, 2006). Cross-national studies suggest that maths teachers from East Asia put more emphasis on the conceptual coherence of teaching (Cai, Ding, & Wang, 2014) and deliver maths lessons in a more coherent manner than their American counterparts (Stigler & Hiebert, 1999; Stigler & Perry, 1988).

Teaching with variation

Like many other teaching theories, teaching with variation is built on various learning theories (Marton & Booth, 1997). At its heart are two types of variation: (1) teaching *concepts* in varying forms (representations) and (2) teaching *procedures* in varying ways (solutions). The idea is that, through teaching with variation, prior knowledge may be consolidated in new situations and new forms, and the essence of new knowledge may be grasped thoroughly in rich and diverse connections with prior knowledge (Gu, 1994). By presenting prior knowledge in varying forms or situations, teachers attempt to scaffold the learning process of new knowledge. Through comparing different solutions to one problem, learners are expected to be able to absorb the essence of specific knowledge and at the same time eliminate misconceptions about it. Teaching with variation is widely accepted and applied by maths teachers from East Asia, such as China (Huang, 2002; Huang, Mok, & Leung, 2006), Korea (Park & Leung, 2006) and Japan (Stigler & Hiebert, 1999).

Teaching with reflection

Reflection is regarded as a crucial way for pedagogical development of teachers. Schön (1983, 1987) proposes three levels of reflection: knowing in action, reflection in action and reflecting on reflection in action. Both research and practice communities have been promoting reflection amongst maths teachers. In initial teacher education, pre-service teachers may improve their reflective skills and form reflective habits through reflecting on lessons delivered by their peers and themselves during placement, which over all may benefit their future teaching practice (Stockero, 2008). Whilst reflection in teaching maths may generate pedagogical change (Walshaw, 2010), well-established reflective communities are claimed to be helpful in cultivating reflective capacities amongst maths teachers (Potari, Sakonidis, Chatzigoula, Manaridi, 2010).

Summary

A dominant theory behind pedagogy in general and maths pedagogy specific to the West is progressive education which offers few options for formal teaching. A beautiful 'dream' behind this is that children could learn best if provided with plenty of opportunities to explore the content independently. Piaget's theory of developmental stages has apparently provided a rationale for ability grouping in schools in England and other countries.

Research in maths pedagogy focuses more on factors within the teachers' mind (i.e. thinking and knowing) than factors without (i.e. teaching). It is thus largely unanswered how to teach the subject matter knowledge well and why. Almost all works on maths pedagogy are either in themselves theoretical or empirical but deeply rooted in certain theories. Empirical studies on maths pedagogy are mostly qualitative in nature, with quantitative data seldom collected.

There were indeed several large-scale studies, but they focused explicitly on *learners'* mathematical thinking (e.g. Hart et al., 1981; Hodgen, Kuchemann, & Brown, 2014). Evidence gathered in these studies was more about *learning* than teaching. Studies that did focus on the teaching process were often limited to a specific teaching factor, such as subject matter knowledge, differentiation, variation and reflection. In particular, studies claiming to investigate teacher PCK often turned out to be explorations of teachers' explanations or classroom dialogue of subject matter knowledge. Subject-related *pedagogy* is still the "missing paradigm" – as Shulman (1986, p. 6) put it. In maths education research, the 'black box' that encloses the connections between teaching and learning is largely undisclosed.

Last but not least, there is also a lack of attempts in seeking international evidence for better pedagogy in mathematics, given the fact that mathematics itself has long gone international as a science, like the discipline of neuroscience – a discipline that has close connections with teaching and learning. Over the past half century, a considerable number of international studies have focused on maths learning outcomes, whereas only a small number of them have stepped into classrooms and systematically investigated the teaching process of the subject cross-nationally. The following section draws upon this part of the literature.

Facts and factors in international maths learning gaps

International comparative studies consistently reveal a performance gap in mathematics between the East and the West (Mullis et al., 2008; Mullis et al., 2004; Mullis et al., 2000; OECD, 2001, 2010). Most recently, the Programme for International Student Assessment, i.e. the PISA 2012 (OECD, 2013) showed that the 15-year-olds from Shanghai, China had been ranked atop the league table in Mathematics, Science and Reading out of 65 countries and economies, while their peers from the UK achieved 26th, 21st and 23rd, respectively, in the three assessed subjects. In mathematics, Shanghai pupils achieved a mean score of 613, whereas British pupils attained 494 on average, lagging behind their peers from Shanghai by 119 points.

The assessment disparity motivated various groups of people, including educational researchers, to find the reasons for the learning gap (e.g. Stevenson & Stigler, 1992). Some examined the social-cultural factors (e.g. Chen & Stevenson, 1995), some investigated the degree of parental involvement (Cai, 2003; Cai, Moyer, & Wang, 1999; Cao, Bishop, & Forgasz, 2007), others compared the curricular content that pupils were expected to grasp (e.g. Ruddock et al., 2008), and some explored the nature of teaching and learning in the classroom (e.g. Schmidt et al., 1996; Stigler & Hiebert, 1999).

In mathematics education and, of course, in many other subjects, the central issues are what should be taught and learned, what is actually taught and learned, and what is attained at the end. The team of the Third International Mathematics and Science Study (TIMSS) divided the educational curriculum into three fundamental dimensions: *the intended curriculum, the implemented curriculum* and *the attained curriculum* (Schmidt et al., 1996, p. 17). Subsequently, *the potentially implemented curriculum* was added in between the original first and second dimensions, making it a four-dimensional conceptual framework(Schmidt et al., 2001).

The intended curriculum consisted of the aims and targets that the pupils were expected to grasp, while *the potentially implemented curriculum* stood for the curricular materials and textbooks. *The implemented curriculum* referred to the process in which the intended and the potentially intended curricula were carried out at the level of schools and particularly the level of classrooms. *The attained curriculum* was what pupils had actually grasped and ultimately achieved in the end as part of their own knowledge. It can be tested through various assessments at the classroom level, school level, national level or international level. Research on teaching is a necessary approach in finding causes for differences of pupil attainment (the attained curriculum), as it is teachers who implement the curriculum in the classroom and provide children *opportunities to learn* the intended curriculum.

With regard to the four categories of curricula, international studies have been done in the comparisons of curricula, textbooks and curricular materials, teaching practices, and learning outcomes, respectively. Given the aim of the study as evaluating teaching practices in mathematics across two countries, China and England, the foci of the following two sections are on international comparisons of *learning outcomes* and *teaching practices,* particularly in mathematics.

International comparisons of maths learning outcomes

Since the first international assessment in the 1950s, there have mainly been three groups of cross-national assessments conducted by three different organizations: the International Association for the Evaluation of Educational Achievement (IEA), the Educational Testing Service (ETS) and the Organisation for Economic Cooperation and Development (OECD). England has participated in most of these assessments, while mainland China has taken part in just three of them: the second International Assessment of Educational Progress (IAEP 2) organised by ETS (Lapointe, Mead, & Askew, 1992) and PISA 2009 and PISA 2012 by OECD (2010, 2013). Nonetheless, several other Chinese-oriented societies, such as Hong Kong, Taiwan and Singapore, have participated in most of the studies, generally ranked in the top five.

The IEA's international studies

The International Association for the Evaluation of Educational Achievement (IEA) was the first organisation to initiate mathematics achievement comparisons across nations. IEA (2012) was founded in 1958 and has subsequently conducted more than 30 research studies comparing pupil achievement not only in mathematics and science but also in other subjects across the world.

The organisation's original motivation was to evaluate both the educational inputs and the educational outcomes. Its primary founders saw the educational world as a natural laboratory that could trigger optimal learning outcomes in children. In order to identify crucial elements that significantly influenced education, various factors were examined in relation to pupil achievement at the international level.

The First International Mathematics Study (FIMS) was conducted between 1963 and 1967, with 13-year-olds and pre-university pupils from 12 countries being comparatively assessed (Husén, 1967). The FIMS found that *the opportunity to learn* was the most significant predictor explaining pupil-achievement differences. In 1980–82, the Second International Mathematics Study (SIMS) was conducted to evaluate pupil mathematics achievement across 20 countries (Robitaille & Garden, 1989). In 1994–95, the Third International Mathematics and Science Study (TIMSS) tested more than 50 million pupils from 46 countries (Mullis et al., 1997). Since 1995, the IEA has been conducting TIMSS in grades four and eight in a four-year-cycle in 1999 (known as TIMSS repeat) (Mullis et al., 2000), 2003 (from 2003 on known as Trends in International Mathematics and Science Study with the same acronym) (Mullis et al., 2004), 2007 (Mullis et al., 2008), and most recently in 2011 (Mullis et al., 2012).

The ETS's IAEP 1 and IAEP 2

As a successful private organisation in the US, the Educational Testing Service (ETS) has, since 1983, managed the National Assessment of Educational Progress (NAEP), which was and still is the largest national organisation in the US assessing pupil progress in a number of subjects over time. Thereafter, ETS has, alongside representative organizations from participating countries, coordinated two international assessments in mathematics and science; entitled International Assessment of Educational Progress (IAEP), they were aimed at testing pupils' knowledge and examining crucial factors influencing pupil performance and attitudes to learning these two subjects.

In 1988, the first study, IAEP 1 (Lapointe, Mead, & Phillips, 1989), assessed 24,000 13-year-olds in mathematics and science in 12 educational systems, and subsequently 175,000 pupils aged nine or 13 from 20 countries were tested in 1990 in IAEP 2 (Lapointe et al., 1992). Of the two assessments, China only participated in the second with 1,650 13-year-olds assessed and ranked in the top place in mathematics in their age group. England took part in both of the assessments ranking 9th out of 12 countries in IAEP 1, and 11th out of 14 countries in the pupil group aged nine and 11th out of 20 countries in the 13-year-old group in IAEP 2.

The OECD's assessments – PISA

From the beginning of the new millennium, the Organisation for Economic Cooperation and Development (OECD) has been conducting a three-yearly survey, the Programme of International Student Assessment (PISA), in both its member countries and partner countries. The aim of PISA is to assess the extent to which 15-year-olds in different countries, at the end of their compulsory education, have acquired sufficient knowledge and have been able to apply such knowledge in their daily life. The assessed subject areas include reading, mathematics and science. England has participated in all four waves of PISA, whereas China has only participated in the latest, PISA 2009 (OECD, 2010) and PISA 2012 (OECD, 2013). Table 2.1 compares the participating and ranking status of the two countries' 15-year-olds.

International comparisons of maths teaching practices

Besides the comparisons of pupil achievement across the world, there are also studies investigating teaching practices across countries. The two IEA's TIMSS Video Studies (Hiebert et al., 2003; Stigler, Gonzales, Kawanaka, Knoll, & Serrano, 1999), the OECD's Teaching and Learning International Survey (TALIS) (OECD, 2009a) and the Learner's Perspective Study (Clarke, Keitel, & Shimizu, 2006) are reviewed.

The IEA's TIMSS Video Studies

In parallel with the international assessment of the Third International Mathematics and Science Study (TIMSS), two video-based lesson studies have been conducted, first in three countries – Germany, Japan and the US in 1995 (Stigler et al., 1999) and second in seven countries – Australia, the Czech Republic, Hong Kong, the Netherlands, Switzerland, Japan and the US in 1999 (Hiebert et al., 2003).

Among various findings, it is worth noting that, in the TIMSS 1999 Video Study (*ibid.*), Japanese pupils spent 74% of their lesson time in making decisions on how to use procedures instead of simply following the teacher's guidance and carrying it out. With regard to the types of problems that teachers asked pupils, Japanese teachers raised a larger proportion (54%) of problems in facilitating pupils to make mathematical connections or investigate mathematical relationships. Nonetheless, teachers in Hong Kong allocated more (84%) problems to pupils to get them familiar with the problem-solving procedures.

Table 2.1 Two countries' mathematics rankings in PISA

PISA	*Shanghai, China*	*UK*
2000 (32 countries/economies)	-	8th
2006 (57 countries/economies)	-	25th
2009 (65 countries/economies)	1st	28th
2012 (65 countries/economies)	1st	26th

Source: OECD

All countries applied a large amount of lesson time to mathematical work, typically through the approach of problem-solving. However, despite surface similarities, there were considerable differences. It was through solving problems that the eighth graders in Japan were introduced to new knowledge. In addition, Japanese lessons also featured more whole-class teaching and longer time per task for pupil seatwork than other countries' lessons. In contrast, Dutch teachers allocated a greater percentage of time for their pupils to work on a set of problems during individual seatwork.

There was also a difference in the emphasis of lessons between countries. Despite the fact that both allocated 76% of lesson time to new knowledge and 24% of lesson time to reviewing prior knowledge, Japanese and Hong Kong lessons had significant differences. Japanese teachers gave more attention to the introduction of new knowledge, whereas Hong Kong teachers focused more on pupils' practice of new knowledge.

Lesson clarity and lesson flow were also taken into consideration. High lesson clarity and low lesson interruption were found in the classrooms of the Czech Republic, whereas the Dutch classrooms were found to have low clarity and frequent lesson interruption.

In addition to the main findings provided by the TIMSS Video Studies' formal reports, there have been a number of follow-up analyses conducted both by members of the TIMSS research team and by many other researchers.

For example, prompted by teachers' understanding of the latest reforms from the TIMSS Video Study and Japanese lesson study, Hiebert and Stigler (2000) proposed a research and practice system for the improvement of mathematics teaching in the US. Nevertheless, what teachers thought could hardly explain the real causal effect relationship behind the legend of top performing countries. It is apparently dangerous to directly *borrow* such understanding from abroad and base an innovation at home on it without proving it works.

In another example, based on the TIMSS 1999 Video Study data, Leung (2005) reanalysed the characteristics of mathematics lessons from Japan and Hong Kong. The secondary analysis found that pupils were exposed to more instructional content in East Asian countries, though they talked less than their peers in the West. In addition to the highest proportion of mathematical language used in mathematics problems, East Asian pupils were given a longer time to solve problems and more opportunities to do proofs, and they were more likely to be engaged in the lessons. Moreover, Hong Kong teachers covered more advanced content in a more coherent manner and with more developed presentations. Nevertheless, how such instructional differences had impacted on East Asian learners' achievement was not further investigated.

There are many other examples where TIMSS Video Studies' data have been either utilised to compare with lessons from countries other than those participating (e.g. Rossella & Alessandra, 2005) or reanalysed afterwards (e.g. Ilonca, 2001; Jacobs et al., 2006; James et al., 2005). However, it is systematic investigations into what worked across countries that was not brought to light in these studies.

The OECD's Teaching and Learning International Survey

To address the issues on the formulation of environments where effective teaching and learning can happen in different countries, the Organisation for Economic Cooperation and Development (OECD) has conducted its first round of Teaching and Learning International Survey (TALIS) (OECD, 2009a) and most recently TALIS 2013 (OECD, 2014). TALIS 2008 was based on data collected from 23 countries regarding "teachers' professional development; teacher beliefs, attitudes and practices; teacher appraisal and feedback; and school leadership" (OECD, 2009a, p. 3). There were 200 schools from each country and 20 teachers from each of the schools participating in the survey. The targeted population consisted of teachers and headteachers at the lower secondary level, i.e. level 2 of the 1997 revision of the International Standard Classification of Education. Teachers and headteachers took part in the survey, filling in two different questionnaires, either on paper or online.

In the TALIS 2008 Teacher Questionnaire (TQ), system-level and pupil-level variables did not draw much attention, whereas school-level and classroom-level variables were systematically surveyed. In the area of teacher beliefs and attitudes, investigations were mainly done around the aspects of beliefs about the nature of teaching and learning, classroom teaching practice, cooperation among staff and job-related attitudes, self-efficacy, job satisfaction and so forth. About teaching practices, questions were asked with regard to time on task, classroom climate, classroom activities, lesson orientation, teachers' knowledge and so on.

In the domain of teacher beliefs, the study found a positive correlation between classroom disciplinary climate and teacher efficacy. In general, teachers across the countries believed that they made a difference. More than 90% of teachers from all countries believed that they made a great difference in their pupils' lives. It was also found that teachers in all countries expressed strong attitudes towards putting great emphasis on structured instructions and pupil-oriented activities. These were more applied by female teachers, whereas male teachers tended more to perceive teaching as the direct transmission of knowledge. In all countries, teachers generally allocated 80% of the lesson time to teaching and learning. However, teacher-pupil relationships varied dramatically across countries, except for Norway where the relationship was well above the international average level and where within-country differences were quite small. In practice, teachers in all countries were likely to set learning targets, make brief revision of previous lessons, give feedback to homework, examine pupils' understanding etc.

In terms of teaching, teacher questionnaires in TALIS 2008 and the subsequent TALIS 2013 focus intensively on teacher self-efficacy and teaching practices, offering insights into what teachers think they are doing across countries. Nonetheless, teachers' self-reported efficacy can only provide one-sided information. The results of TALIS survey could have provoked even deeper thoughts if there were extra data collected through other methods, such as classroom observations which would help reveal what teachers were actually doing in contrast with what they claimed to be doing in the survey, focus groups which could lead to interactions between teachers on what counted as more effective lessons, and so on.

The learner's perspective study

Through international collaboration, Clarke et al. (2006) conducted the Learner's Perspective Study (LPS) with researchers from 12 countries: Australia, China, the Czech Republic, Germany, Israel, Japan, Korea, the Philippines, Singapore, South Africa, Sweden and the US. Each country team chose three teachers as competent teachers on local standards. Given the means of selecting teachers, it was not LPS researchers' intention to sample representative teachers and typical lessons in each country for cross-national comparisons.

The research design was qualitative in nature. For every teacher, a sequence of 10 consecutive lessons taught by each teacher participant was observed and video-recorded. For every classroom, three questionnaires were given to the teacher, and a post-lesson interview with the teacher and another with two focus pupils were conducted to collect their accounts on lesson events, using lesson videos as stimuli.

The reason LPS researchers did not intend to collect national samples and insisted on collecting in-depth qualitative data was a result of their critical reviews on the research methods and findings of the two TIMSS Video Studies. LPS researchers (Clarke et al., 2007) criticised the points that TIMSS Video Study researchers held regarding the focus of comparisons and the levels at which variables should be chosen and compared. They attempted to test the extent to which the national teaching patterns their TIMSS counterparts found would work on the LPS data from the US, Germany and Japan. Then, they employed the TIMSS Video Study lesson codes generated from lessons in the three countries and applied them to the LPS data set of the three countries. These codes were claimed not to be identifiable in almost each LPS lesson, and no evidence suggested the existence of the national patterns that TIMSS researchers found. Conversely, there is a strong argument in the LPS report that lessons varied considerably within and between individual teachers' classrooms in the three countries (Clarke et al., 2007). They thus argued that, in international research on teaching, it would be more sensible to compare lesson events than to compare lesson patterns/scripts on which the two TIMSS Video Studies were focused.

Moreover, the TIMSS Video Study researchers (Givvin, Hiebert, Jacobs, Hollingsworth, & Gallimore, 2005) reanalysed the TIMSS 1999 Video Study data and characterised each country's lessons in three dimensions: Purpose, Classroom Interaction and Content Activity. This aroused further debates from the LPS researchers over the undue simplicity of the TIMSS Video Study researchers' approaches. They argued that the TIMSS Video Studies' findings were based on "over-inclusive codes" (Clarke et al., 2007, p. 283) which increased the international adaptability of the coding scheme but decreased its discrimination in reflecting inter-class variation. Thus, the LPS researchers decided to focus on "the location of the lessons in the classroom of competent teachers" and "learning outcomes provided in post-lesson interviews" (*ibid.*, p. 281). The former led them to observe a sequence of each teacher's lessons on a topic rather than one lesson and interpret connections between lesson events and teaching goals. The latter was collected in video-stimulated interviews with focus pupils.

Teaching studies conducted by the two teams of researchers on both sides have their strengths and weaknesses. TIMSS Video Study researchers tried to see things holistically, whereas LPS researchers scrutinised lessons too closely, therefore losing the opportunity to capture bigger pictures. Other TIMSS Video Studies' strengths and weaknesses have been discussed formerly. For the Learner's Perspective Study, its advantages lay in the application of innovative methods in research on mathematics teaching and learning and the rich accounts of the insiders' (i.e. teachers and pupils') thoughts. However, such a heavy emphasis on the insiders' views prevented research findings from being thoroughly synthesised among countries. There was a severe lack of quantitative measurements to supplement the multiple perspectives collected, which made the study less scientific than it could have been.

Chapter conclusion

This chapter gives a review of (1) typical pedagogical theories and practices in school mathematics and (2) large-scale studies conducted in two domains of international comparisons – the comparison of mathematics learning outcomes and the comparison of mathematics teaching practices. The two domains happen to be the two fundamental variables in teaching effectiveness research which will be reviewed in the next chapter.

Works on mathematics pedagogy are either theories in themselves or deeply rooted in theories. Empirical efforts are generally devoted to the learning of mathematics and teacher subject matter knowledge. Evidence about what works in the teaching of mathematics is still fragmented and somehow recondite.

The international mathematics assessments organised by the IEA, the ETS and the OECD over the years consistently revealed that pupils from a Confucian-heritage-cultural (CHC) background (including Taiwan, Hong Kong, and mainland China) steadily outperformed their peers in the West. Although pupils from mainland China only participated in three of those assessments, they performed the best and were ranked at the top on all occasions (Lapointe et al., 1992; OECD, 2010, 2013).

The number of international studies on the teaching of mathematics is relatively small compared with that on pupil performance in the subject and very few of the large-scale comparisons reviewed involved teachers from mainland China. There is clearly a dearth of cross-cultural studies on maths teaching involving China, particularly given the results that Chinese pupils topped the international league tables every time they participated. Moreover, few of international comparative studies have attempted to measure and correlate learning outcomes and teaching practices in one study.

Notably, from the very beginning of teaching effectiveness research, the impact of teaching on learning has always been the main research interest of researchers in this field. The following chapter reflects upon the evolution of four decades of educational effectiveness research (EER) and its core subfield – teaching effectiveness research (TER), synthesising key findings of TER studies carried out in different times and spaces.

3 Four decades of educational effectiveness research in the West

Chapter overview

- Educational effectiveness research and its subfields
- Early teaching effectiveness research in the US
- Teaching effectiveness research in the UK
- Teaching effectiveness research in other nations

This chapter starts from a brief reflection upon the evolution of EER, traces the development of its subfield TER to its 'childhood' in the US and then in the UK and arrives at national studies dotted across various countries in the West. At the end of this chapter, the conclusion highlights gaps in the reviewed literature and the relevance of this study to both research and practice.

Educational effectiveness research and its subfields

One of the earliest areas in EER is the School Effectiveness Research (SER) which historically had its 'first' voice in responding to Coleman et al.'s (1966) assertion that schools made no difference and the subsequent long-term defence by many other researchers regarding the actual effects of schools and the functions of the educational system, either outside or inside the schools. As follows, the developmental stages of EER will first be traced; second, there will be a brief overview of the main characteristics of the field; third, evidence from previous studies will be presented to explain the difference between teacher effects and school effects.

Five phases of educational effectiveness research

Based on many previous reviews of the history of EER (e.g. Creemers, Kyriakides, & Sammons, 2010; Mortimore, 1991; Reynolds, 2010; Sammons, 1999; Teddlie & Reynolds, 2000), the development of this field has moved through five phases (Reynolds et al., 2011).

The first phase of EER started as a reaction to the Coleman Report (Coleman et al., 1966), as mentioned before. Then followed the empirical studies conducted by Edmonds (1979), Rutter, Maughan, Mortimore, and Ouston (1979) and Smith and Tomlinson (1989) and small-scale case studies carried out by Weber (1971) and Reynolds (1976).

The second phase occurred from the mid-1980s when multilevel methodologies (Goldstein, 2003) and other studies applying such methodologies emerged and consistent educational effects in certain areas in the long-term proved the scientific properties of school effectiveness.

The third phase came in the early to mid-1990s as numerous researchers attempted to explore the reasons that caused schools to perform differently in educating children. In the US, there was the Louisiana School Effectiveness study conducted by Teddlie and Stringfield (1993), and in the UK, Sammons, Thomas, and Mortimore (1997) researched into the effects of departments on pupil performance and the effectiveness of schools. Such actions were also realised through a number of prominent reviews of the field (Creemers, 1994; Reynolds et al., 1996; Scheerens & Bosker, 1997; Teddlie & Reynolds, 2000; Teddlie & Stringfield, 1993).

The fourth phase started in the mid-1990s and went through a decade with its influence and development remaining today. The remarkable theme in this phase is the internationalisation of the EER field, and the close connections and collaborations of school effectiveness (SE) researchers with school improvement (SI) practitioners. The seeds of EER have travelled from a few initial countries to many others, such as the Dutch study by Bosker and van der Velden (1989), the French study by Grisay (1996), the Australian study by Hill and Rowe (1996), Flanders' studies by Van Damme et al. (2006), De Fraine, Van Damme, and Onghena (2007) and Verachtert, Onghena, and Ghesquiere (2009). The wider geographical development of EER also formed the trend of internationally joint research in the field. Within the field, SE and SI researchers are becoming more comfortable with each other's methodologies in that SE's quantitative perspective gets in line with SI's qualitative vision and the two collaboratively investigate educational issues through mixed-methods.

The fifth phase started at the end of 2000s and is currently still rapidly developing. The focus of EER in this phase mainly addressed the dynamic, rather than static, characteristics of relationships between all the effective components and the impact of the interactions between these components on educational outcomes (Creemers & Kyriakides, 2008).

Trends in EER

In EER, educational outcomes are increasingly seen as being formulated by various types of factors working collectively at multiple levels – the pupil, the classroom, the school and the system levels (Creemers & Kyriakides, 2006; 2008).

Although the prevalent criterion in EER is still pupil academic performance, there are also gradually more and more researchers who pay attention to other aspects of educational outcomes, such as pupil well-being (Konu, Lintonen, & Autio, 2002; Van Landeghem, Van Damme, Opdenakker, De Fraine, & Onghena, 2002) and progress drive (Van de gaer, De Fraine, Van Damme, De Munter, & Onghena, 2009; Van der Werf, Opdenakker, & Kuyper, 2008). In addition, pupils' learning outcomes are gradually becoming more researched over a longer time span, which thus helps judge how long the educational effects can last and to what degree such effects can follow the children through their later schooling or their life after leaving schools.

Children are organised in classrooms which are nested in schools and then systems. EER researchers have long been concerned with the question as to which educational level makes a bigger difference. Thus, effect sizes of different levels have been compared either through empirical investigation or theoretical meta-analysis. The most scrutinised pair is schools versus teachers – which makes a bigger difference? In the following section, we will see research evidence on this accumulated so far.

Teacher effects vs school effects

The effect size of teaching has consistently been found in various countries to be much bigger than that of schools or systems. Through a secondary analysis of the results from the Second International Mathematics Study (SIMS), Scheerens, Vermeulen, and Pelgrum (1989) found the degree of educational impact on pupil achievement varied considerably from the classroom level to the school level. Across eight countries, schools explained 0% to 9% of the variance of pupil achievement, whereas classrooms (teachers) explained 16% to 45% (*ibid.*, p. 794). Investigating the teaching and schooling of both primary and secondary literacy and numeracy, the Victorian Quality Schools project (Hill & Rowe, 1995) examined the educational effects on 13,700 pupils from 99 Australian primary and secondary schools. The results suggest that, from primary to secondary and from literacy to numeracy, teacher effects are approximately 5 to 10 times as large as school effects. In the 1996 Elementary School Climate Study (Willms, 2000), considerably bigger differences in pupil performance were found between classes than between schools and between districts in Canada. The meta-review of 500,000 studies by Hattie (2003) from New Zealand again indicated the essential role of teachers as a main source of pupil learning improvement. It suggested that 30% of the achievement variance was explained at the teacher level, whereas schools counted for 5% to 10% of the variance. Another meta-analysis done by Luyten (2003) reviewed a number of empirical studies and found bigger teacher effects than school effects particularly in primary schools in the subject of maths. The reviews done by Reynolds and Teddlie (2000), and Scheerens and Bosker (1997) also came up with the same finding that teacher effects are much larger than school effects.

Judging by the research evidence so far, it is reasonable to say that, whilst schools matter, teachers make a much bigger difference in children's education. After a brief review of the evolution of EER and keys issues in its subfields SER and most specifically TER, we could see that if we wanted to make a greater difference in children's maths learning, we would need to pull the lever of classroom teaching. Researchers in EER have always been putting their interests in and focus on classroom teaching and learning ever since the field was born. This is why we are now starting our review of TER literature by heading back to the very beginning of the field in the US four decades ago.

Early teaching effectiveness research in the US

In the early 1970s, the development of human behaviourism theory contributed to and motivated research towards a focus on the effectiveness of teaching, according to the developmental process of children's cognitions and the corresponding learning procedures in their minds. Effective teachers are the ones who, at the least, are able to successfully improve pupils' learning outcomes in the subject matter. In some sense, teaching and learning are regarded as a cause-and-effect pair. Since the core aim of teaching is to enhance pupil learning in specific subjects, it is essential to understand how learning is processed in children's minds before one starts to consider the most effective way of teaching.

Like many of their contemporaries, Rosenshine and Stevens (1986) linked the process of teaching to research on 'human information processing' (p. 378) so as to understand the ways in which teaching worked. According to information processing theory, human brains are limited to and are only able to effectively cope with a certain amount of information each time (Becker, 1978; Miller, 1956). Such a feature of human learning suggests that new knowledge needs to be broken into small segments and taught in small steps (Rosenshine & Stevens, 1986). Furthermore, new knowledge cannot be grasped and stored in a stable manner unless learners keep practising it to an overlearning degree so that it can be transferred from temporary memory to long-term memory, providing sufficient available space for more knowledge to come in (Gagné, 1985; LaBerge & Samuels, 1974). In different subjects, various layers of knowledge are constructed in different ways. Overlearning words enables pupils to pay more attention to the comprehension of reading materials, while overlearning basic skills, such as calculations in mathematics, enables pupils to fully concentrate on high-cognitive-level work, such as problem-solving and mathematical reasoning.

Building on the principles of human information processing and the effective models developed by other researchers (Gagné, 1970; Hunter & Russell, 1981), Rosenshine and Stevens (1986, p. 379) suggested a general model of effective instruction which contained six components: (1) daily review and checking homework, (2) presentation, (3) guided practice, (4) correctives and feedback, (5) independent practice and (6) weekly and monthly reviews. This structure echoed with different effective teaching models identified by other researchers (Good & Grouws, 1979) and showed its consistency with effective teacher behaviours found in other studies (e.g. Fortune, 1967; Soar, 1977). These models commonly emphasised the necessity of revising prior knowledge before inputting new knowledge and the need to develop teaching steps according to the internal structure of various cognitive levels of knowledge.

Around the 1970s and 1980s, there were hundreds of studies done on teaching effectiveness in the US. Researchers usually did pre- and post-tests of pupil academic performance and then linked the differences between pupil test results over time to teacher behaviours which were mainly measured through classroom observations or questionnaires. Many studies consistently found positive effects of certain teacher behaviours on pupil achievement. This section of literature review is therefore aimed at seeking the most effective teacher behaviours which were identified and building connections between those behaviours.

As follows, first, the key definitions of teaching effectiveness featured in the literature will be retrieved; second, the developmental process of research methods and theoretical frames in teacher effectiveness studies in the 1970s of America will be briefly reviewed; and then the main discussion will address the most identified effective teacher behaviours. These behaviours include: (1) *profound subject matter knowledge*, (2) *academic-oriented instruction*, (3) *providing a high opportunity to learn*, (4) *more time allocated to direct instruction*, (5) *clear and structured presentation*, (6) *maintaining a brisk lesson pace*, (7) *prompting a moderate to high success rate*, (8) *active whole-class interaction*, (9) *proper use of teacher questioning*, (10) *appropriate praise and criticism*, (11) *assigning homework regularly with timely and clear feedback* and (12) *efficient classroom management*. Finally, the identification of gaps in the literature will be presented and discussed as a conclusion to this part of the literature review.

The definition of teaching effectiveness

Teachers' teaching at the classroom level is seen as a process towards gaining certain expected products, i.e. learning outcomes. The more learning outcomes produced, the more effective the instruction (process) is. Research on the connections between the process and the product was called *process-product research* in the 1970s US. Based on such ideas of effective teaching, earlier American teaching effectiveness researchers carried out a huge amount of classroom studies, mainly through lesson observations, correlating teacher behaviours with learning outcomes/gains (Brophy & Good, 1986).

The methodological and conceptual development

Prior to the 1970s, there was a lack of "systematic research linking teacher behaviour to pupil achievement", and the existence of such a research gap was partly attributed to the fact that the cost of observations at that time was considerably high and also partly caused by previous concepts on teaching (*ibid.*, p. 329).

In the era before the 1970s, researchers had already started seeking factors in teachers that would increase the quality of teaching and learning, but they were more interested in connecting pupils' achievement with teachers' personal traits than with specific teacher behaviours. Thus, Brophy and Good argued that the "early concern with teachers' personal traits led to presage-product rather than process- product studies" (1986, p. 329).

Thereafter, research forms were limited at quasi-experimental comparisons of the achievement of pupils taught in one way of teaching with the achievement of those taught through another teaching approach. There were just a few studies conducted in this domain (*ibid.*) among which apparently little difference was identified regarding pupil achievement between classes taught with different methods.

The systematic observation of teacher behaviours was triggered by the educational trend in facilitating good classroom climates and effective teaching in the 1950s and 1960s, and some modest findings were achieved (e.g. Flanders & Simon, 1969; Gage, 1965). However, at the same time, there was a concept calling for a shift of the research focus from teachers to the curricula, i.e. *what* to teach, on which some researchers put more emphasis than on *how* to teach (e.g. Walker & Schaffarzick, 1974). Moreover, early school effectiveness research also emphasised the contribution of teachers less than that of schools to the improvement of pupils' achievement.

Nevertheless, during the 1960s, significant progress in research conceptions and methodologies had also been made and gradually proved the process-product approach to be a valid way of researching educational effectiveness. New observation systems had been developed with new process variables identified. The number of observation systems had dramatically increased up to more than a hundred by the year 1970. Meanwhile, research had identified steady relations between particular teacher behaviours and pupils' learning outcomes/gains (Rosenshine, 1971). Such correlations were predicted to possibly be working in nonlinear ways and dependent on differences of pupil ability and background. In addition, Rosenshine (*ibid.*) pointed out that several key factors, such as contexts, the subject matter, grade levels and terminologies applied in the measurement, might have a significant influence on research findings in this area. From the beginning of the 1970s up to the early 1980s, a great number of studies had been conducted and new variables examined in the area of process-product research on teaching and its impact on learning.

Overall, there were some events that contributed to the continuous development of research on teaching during the 1970s. Rosenshine and Furst (1973) called for the utilisation of direct observations in the area of teaching studies as well as systematic research on the descriptive-correlational-experimental loop to describe lessons with quantitative instruments, correlate observed variables with pupil achievement and examine corresponding causes of specific effects in the correlational connection. Moreover, Dunkin and Biddle (1974) defined research on teaching and made distinctions between teaching studies and research in other educational areas. They also critically stressed the importance of evidence-based policymaking, argued the inappropriateness of making educational prescriptions based on untested theories and pointed out the necessity of prescribing education based on reliable empirical findings. In addition, they emphasised the importance of contextual variables in research on teaching. Last but not least, the American federal agencies, such as the Office of Education and the National Institute of Education, recognised the importance of research on teaching and provided sufficient funding for numerous large-scale studies on the effectiveness of teaching, which provided a positive national climate for teaching effectiveness research (TER) to flourish.

Effective teacher behaviours identified

The great number of TER studies makes it feasible that the most frequently observed and identified effective teacher behaviours could be located and organised for further analysis and synthesis.

Profound subject matter knowledge

As research indicated, effective teachers generally demonstrated stronger subject knowledge and clearer understanding of how to put it into well-structured instructional practices than did less effective teachers (Tikunoff, Berliner, & Rist, 1975). This was consistent with the recognition of the importance of teachers' pedagogical content knowledge in Schulman's seminal work (Shulman, 1986, 1987) and in line with Ma's (1999) findings that the gap in mathematics learning between pupils from two countries was partly attributable to Chinese primary teachers' more profound understanding of the subject matter than their counterparts in the US.

Academic-oriented instruction

More effective teachers were found to spend more time on academic activities and successfully facilitate pupils to engage in such activities longer than less effective teachers (Fisher et al., 1980). They were also good at briskly switching between lesson segments, therefore spending less time on transitions and discipline maintenance than their less effective peers (*ibid.*). Teachers who produced the largest learning gains devoted the maximum proportion of time to active academic teaching and seatwork supervision and the minimum proportion of time to managing pupil behaviours (Berliner & Tikunoff, 1976, 1977). Effective teachers were also good at tailoring appropriate subject content and academic tasks for pupils according to their current knowledge levels and readiness for new knowledge (Fisher et al., 1980).

In the Follow Through Evaluation Study, Stallings and Kaskowitz (1974) studied lessons in 108 Grade-1 and 58 Grade-3 classes taught by experienced teachers who were implementing one of the seven Follow-Through models. Data were collected regarding the amount of time allocated to various activities and the frequencies of teacher-pupil interactions. The results indicated that the teaching components derived in the Follow-Through models had a positive impact on pupils' standardised test scores. Overall, the pupils that gained more were the ones whose teachers used more time in active instruction of subject content and provided sufficient supervision to pupils during individual work when necessary. On the contrary, pupils that gained less were the ones whose time was largely spent on non-academic activities and whose work was completed independently with little supervision. Given the pupil attainment level before the study as below average, the findings also suggested that low-ability pupils might need more frequent instruction from their teachers so as to improve academic performance.

The California Early Childhood Education programme conducted by Stalling, Cory, Fairweather, and Needels (1977) also obtained similar findings which coincided with those from the Follow Through Evaluation Study in that the quantity of teachers' academic instruction played an essential role in improving pupils' achievement.

Another study by Stallings and her colleagues (Stallings, Cory, Fairweather, & Needles, 1978), on reading instruction in 43 secondary schools, again identified the extent of academic-oriented instruction as a key factor correlating with pupil academic performance. Negative factors included little teacher-pupil interaction, teacher classroom management (instead of academic instruction), pupil independent work without sufficient supervision and so forth.

Providing a high opportunity to learn

In classrooms where effective teaching could happen, besides sufficient orientation of teaching in academic content, another crucial factor is providing the highest opportunity for pupils to learn (Stallings et al., 1977).

The teacher, as the initiator of such an opportunity, should not only ensure a large enough *quantity* of learning, such as a high coverage of curricular content, pupils' more time on task etc., but also guarantee a high *quality* of such academic engagement. Pupils should be thoroughly engaged into the thinking and learning of the subject knowledge. Thus, an ideal opportunity to learn consists of two mutually dependent dimensions: *time allocation* and *academic engagement*, which cannot create change independently and can only work collaboratively on the improvement of learning (Arehart, 1979; Borg, 1980; Crawford et al., 1978; Fisher et al., 1980). In general, teachers delivering high-quality lessons tend to leave plenty of time for teacher-directed instructions, allocate less time to individual seatwork, and maintain equal supervisions to every pupil during seatwork, particularly at the primary phase.

More time allocated to direct instruction

Initially motivated by the interest in observing the amount of teacher talk, Flanders and his colleagues designed an observational instrument, the Flanders Interaction Analysis Categories (FIAC) (Flanders, 1970), in order to describe directness and indirectness of teaching in the classroom. Coding teacher behaviour with 10 behavioural categories in the FIAC system, the researchers found teacher talk occupied two-thirds of classroom conversations and had positive effects on pupils' achievement. Moreover, indirect teaching was found to have more positive correlation with pupils' attitudes than with their achievement. Direct teaching was found to be more appropriate for teaching facts or skills, whist indirect teaching was found to be more suitable for teaching abstract reasoning or creativity. Since basic skills and facts were mainly addressed at the primary phase, direct teaching was much more appropriate for this stage of learning than was indirect teaching. Conversely, indirect instruction was much more suitable for nurturing creative thinking at the secondary phase. Teacher-directed teaching might not, however, always work effectively in improving pupils' achievement without clear presentations and other effective elements to complement it.

Clear and structured presentations

In implementing a session, effective teachers were found to be more able to plan ahead and then present their teaching in a clearer and more structured way than others. Such clarity is mainly evident in two domains: teacher expressions and lesson structures.

A strand of research interest grew around the clarity of teacher presentations during lessons, with the main negative factors being vagueness of terms, digressions, discontinuity and saying "uh". Smith and Land (1981) found pupils' achievement scores were influenced and reduced when vague terms were added into well-planned lessons. In addition, in three out of four studies, they also discovered similar decreases of pupil academic performance when clear presentations had been mixed with digressions, for example, when teachers repeated certain words or added 'I mean' in between words of a sentence which was originally planned to be fluent. Evidence from Kounin's study (1970) proved that lesson momentum could be interrupted by any discontinuity delivered by teachers, for example, in mentioning irrelevant content or introducing relevant content at an unsuitable time. Similarly, Smith and Cotten's (1980) study suggested that a high level of content changes made to the original clearer presentation led to the reduction of pupil achievement. Teachers with a high level of teaching clarity were able to generate higher pupil achievement (e.g. Land, 1979; McCaleb & White, 1980).

Apart from clarity in teacher expressions, some studies also focused on the correlations between the clarity of lesson structures and pupil learning outcomes. Applying both low-inference and high-inference ratings, Fortune (1967) observed and analysed trainee teachers' presentational skills in teaching English, mathematics, or social studies in upper primary classrooms. The findings clearly distinguished more effective teachers from less effective ones in that the former were more likely to start lessons with an overview or analogy, and reviewed old knowledge when necessary.

In the Missouri Mathematics Effectiveness Project (Good and Grouws, 1979) – a treatment study, an instructional model was introduced to teacher participants. This model consisted of five main domains: daily review, development, seatwork, homework assignment and frequent reviewing, with more emphasis on active whole-class teaching. The implementation of the instructional structure with clarity proved to significantly improve pupils' achievement over several months in the classes taught with the model. Clark et al. (1979) also found a positive correlation between clearly structured lessons and an improvement in pupils' achievement. Furthermore, clear transitional signals between the components of lessons were also regarded as crucial in making lessons well-structured and therefore more effective.

Overall, clearer lesson structures coincided with the information-process theory in building connections between old and new knowledge and also between short-term and long-term memories by mapping revision, explanation, demonstration and assigning classwork and homework properly throughout the lesson. This means that the clarity of lesson structure has its effectiveness rooted in both empirical and theoretical areas.

Maintaining a brisk lesson pace

The speed at which teachers teach varies considerably across lessons and also makes the levels of teachers' effectiveness significantly different from one another.

A longitudinal large-sample study, the Texas Teacher Effectiveness Study (Brophy & Evertson, 1976) followed 165 teachers' teaching through four consecutive years, paying attention to the correlations between teaching pace and pupil learning outcomes. It was found that pupils learned best when their teachers broke learning tasks into small and easy steps and then taught them at a brisk pace. Moreover, some differences were also identified between low-SES and high-SES pupils in terms of the corresponding optimal teaching pace and how small the steps should be. Low-SES pupils should be taught in smaller steps with certain repetitions when necessary, while high-SES pupils could benefit more from a brisker instructional pace and in slightly larger steps.

In the Stability Analyses Study, Good and Grouws (1975) collected data from lessons given by 103 third and fourth grade teachers in two consecutive years and observed the stability of teaching effects on pupil attitudes and achievement in all subjects. The results were statistically significant but at low levels. They then decided to focus on one subject, mathematics, by comparing the teaching effects of nine effective teachers with those of nine less effective teachers on their pupils' achievement. In a later study, the Fourth Grade Naturalistic Study, with an extra group of 23 fourth grade teachers added to the original 18 teachers, Good, Grouws, and Beckerman (1978) discovered that more effective teachers taught the curricular content at a brisker pace than their peers.

Overall, a brisk pace saves teachers' teaching time to address more new knowledge, and consequently exposes their pupils to more academic content in a session, which significantly increases their opportunity to learn (Good, Grouws, & Ebmeier, 1983; Griffin & Barnes, 1986; Lampert, 1988).

Prompting a moderate to high success rate

In addition to a brisk pace, collecting data from 25 Grade 2 and 21 Grade 5 classes, researchers from the Beginning Teacher Evaluation Study (Berliner, Fisher, Filby, Marliave, 1978; Fisher et al., 1978) also found that success rates were positively correlated with pupils' learning outcomes. For the whole sample, on average, pupils achieved 50% high success, about 50% medium success and 0–5% low success. Such effects varied slightly across different pupil age groups. The learning of younger pupils was promoted more by high success rates than that of older ones for whom a moderate level success rate was appropriate. For example, Grade 5 pupils averaged 35% high success rates. Overall, it was found that pupil achievement was positively correlated with approximately 65–75% high success, 25–35% medium success and 0% of low success.

Active whole class interaction

Besides the effective behaviours discussed above, in a holistic view, effective teachers were found to be teaching more actively at the whole-class level and thus striving towards a high level of pupil engagement.

Good and Grouws (1979) conducted process-product research along with their colleagues in the 1970s and developed a mathematics teaching model called active whole-class instruction in mathematics. They found the teachers who implemented whole-class instruction could be either effective teachers or ineffective teachers. Whole-class instruction, they noted, would only add value to the effectiveness of teaching if teachers implemented it in a more active way, with efficient classroom management and appropriate teaching skills. Likewise, in the Junior High Study, Emmer, Evertson, and Brophy (1979) found that more effective teachers allocated more time to whole-class teaching and demonstration but less time to seatwork, taught more actively and aroused more class-level discussion.

In general, effective teachers conduct more interaction with the whole class rather than with individuals with whom they only interact in a brief but efficient way. Supervision to each pupil during the seatwork stage is evenly distributed. Effective teachers attempt to track their pupils closely in the whole class scope and ensure their thorough comprehension of the knowledge before sending them to seatwork in which re-teaching is not necessary and therefore less time is needed for completing the practical work than in less effective teachers' classes (Evertson, Anderson, & Brophy, 1980).

Proper use of teacher questioning

Abundant evidence also consistently confirms the importance of teachers asking appropriate questions for improving learning outcomes.

More effective teachers are less likely than other teachers to ask questions to which pupils are not able to provide correct answers (Good & Grouws, 1977). On the other hand, there is also evidence for the curvilinear relationships between pupils' academic performance and percentages of correct answers they make regarding teachers' questions. Brophy's (1973) large-scale study found different optimal levels of correct-answer percentages for high-SES and low-SES pupils: the former progressed best when around 70% of their answers were correct; the latter did best when about 80% of their answers were correct. The types of questions teachers asked therefore mattered. Effective teachers tended to ask more direct factual questions so that their pupils were more able to respond quickly with correct answers. In the Follow Through Evaluation study (Stallings & Kaskowitz, 1974), teachers who utilised more close-ended questions were found posing greater positive effects on pupil academic performance than those who applied more open-ended questions.

More effective teachers ask their pupils more questions than teachers with lower effects (Emmer et al., 1979; Evertson et al., 1980). There are also positive correlations between the quantity of process questions and the level of pupil achievement. Teachers asking more process questions than product questions provide an essential opportunity for pupils to clarify their thinking in the class, which at the same time makes their thinking more observable to the whole class and the teacher. The teacher is then able to diagnose more accurately the level of pupils' understanding of certain points of knowledge and timely prescribe pupils with optimal 'treatment'.

Appropriate praise and criticism

More effective teachers are good at maintaining a balance between praising and criticising pupils when confirming or denying their academic or conduct performance. The appropriateness of praise or criticism is among the factors most investigated in the studies that fall in this historical part of the literature.

In reacting to pupil responses or giving seatwork or homework feedback, teacher praise at a moderate level generally generates a positive effect on pupil attitudes but mainly produces a positive achievement change in low-SES classes; it is somehow irrelevant to or even shows a negative connection with pupil achievement in high-SES classes (Brophy, 1973). Nevertheless, teacher praise is unrelated to learning improvement when such praise is about pupils' good conduct. In addition, there are negative correlations between teacher criticisms or punishment about pupils' misconduct and learning gains. When coping with misconduct, effective teachers tend to simply react with commands or warnings instead of criticism or punishment which less effective teachers usually employ.

In responding to pupils' wrong answers, effective teachers tend to make brisk and clear judgement without hurting personal dignity and then gradually give clues to pupils so as to enable them to answer the question more independently (Anderson, Evertson, & Brophy, 1979). In addition, in summarising the incorrectly-answered questions, effective teachers tend to explain the process rather than simply providing the correct answers, hence deepening pupils' understanding and lowering the possibility of their future repetition of similar mistakes. With the theory of the inverted-U curvilinear process-outcome relationships, Soar and Soar conducted five process-outcome studies in 55 classrooms of Grades 3–6 (Soar, 1966), 20 Grade 1 classrooms (Soar & Soar, 1972), 59 Grade 5 classrooms (Soar & Soar, 1973; Soar & Soar, 1978), 22 Grade 1 classrooms (Soar & Soar, 1973; Soar & Soar, 1978), and 289 primary grades classrooms (Soar, 1973). Six observational systems were applied in the five studies in which negative climates, such as teacher criticism and negative emotions, were found negatively correlating with pupils' performance, and positive climates, such as teacher praise and positive emotions, were found having no significant positive relation with pupils' achievement. Moreover, for low-SES pupils, positive effects become more positive and negative effects more negative.

Assigning homework regularly with timely and clear feedback

Effective teachers tend to assign, evaluate and provide feedback to homework more frequently than their less effective counterparts (Good & Grouws, 1979). Starting a lesson with a review of the previous lesson, effective teachers are likely to process such a review by giving feedback about the previous homework to the whole class and thus reconstruct the prior knowledge in a class-level discussion before introducing new content, which is in line with the information-process theory (Becker, 1977; Emmer, Evertson, Sanford, & Clements, 1982; Good & Grouws, 1979; Reid, 1978–1982). However, the positive effect of homework cannot happen without the functioning of the complete set of teaching factors. Judging by the fact that pupils are already capable of dealing with the seatwork content similar to the intended homework, an effective teacher is able to anticipate pupils' readiness for the homework before assigning it (Good & Grouws, 1979).

Efficient classroom management

Effective teachers are good at maintaining the teaching and learning process in order, allocating time reasonably to various activities, interacting properly with pupils and building a positive relationship with them.

More effective teachers are able to manage the class more efficiently because they are better at identifying and dealing with problems immediately or preventing them from happening. They are more focused on the academic targets and therefore allocate more (ideally all) time to academic activities than less effective teachers. For instance, in the Fourth Grade Naturalistic Study, Good and Grouws (1977) found that effective teachers could manage their classes better than their less effective colleagues even when the class size was larger. Furthermore, effective teachers spend less time either on the transition between lesson segments or in maintaining classroom order. Effective teachers also spend less time dealing with previous assignments, because their pupils have already grasped the knowledge from the previous lesson(s) thoroughly and firmly.

Moreover, effective teachers paced forwards through the textbook more briskly, covering 1.13 pages per day on average, while their peers with lower efficacy could only cover 0.71 pages every day in the study by Good et al. (1978). Consequently, it was summarised that the brisk pace in effective teachers' lessons was formulated by first starting with clearer presentations, second by knitting together explanations and questions without densely utilising teacher monologue, third by asking factual questions and promoting immediate correct responses and fourth by explaining the process rather than simply giving correct answers when reacting to incorrect responses (*ibid.*).

Furthermore, Good and Grouws (1979) found that, in addition to less time on homework revision, there was less time spent on seatwork in effective teachers' lessons. Effective teachers were found good at allocating their attention evenly to each individual and sufficiently supervising every pupil at seatwork, while less effective teachers tended to leave their pupils to independent work without requisite guidance (*ibid.*).

Effective teachers have more sense of professional responsibility and are more willing to interact with pupils (Brophy, 1973). They are more likely to build a business-like instead of personal relationship with pupils, are keener on their work and interact with their pupils more politely and happily than ineffective teachers.

Conclusion

TER studies, initiated during the 1970s in the US, form a profound knowledge base for succeeding researchers home and abroad with which to advance the field. The review of this part of literature sets the historical and theoretical background for the study and suggests certain gaps. With regard to the research approaches, the gaps are mainly: (1) quantitative methods dominated the research convention and qualitative methods were rarely utilised; (2) there is a lack of teacher voices in what works; (3) there is a lack of international TER evidence, let alone international dialogue on the effectiveness of teaching. Moreover, as Rosenshine and Furst (1973) called for in the early 1970s, there is still a lack of understanding on the extent to which effective teaching is bounded by certain factors, such as grade levels and subjects. Models of effective teaching for specific schooling phases and in various subjects, for example mathematics, are still fuzzy, awaiting more recent systematic studies to push the research boundaries and inform research, practice and policy with updated evidence.

Next, we will follow the footsteps of TER from the US to the UK and see several major large-scale TER studies and notable effective teacher behaviours identified there.

Teaching effectiveness research in the UK

By the end of the 1960s, British education had been much influenced by the then popular idea of child-centred classrooms which were recommended by the 1967 primary review known informally as the Plowden Report (Central

Advisory Council for Education, 1967). The idea was also strongly influenced by Piaget's (1962) child cognitive developmental theory published during the same decade.

Around the same period, the concept of process-product research on teaching effectiveness was imported from the US to the UK. British researchers carried out a number of studies on teaching effectiveness (e.g. see in Reynolds, 1985) among which were historically six major large-scale studies investigating the effectiveness of teaching, applying the process-product research concept and collecting and analysing large amounts of data over time. In chronological order, the first large-scale teaching effectiveness study was the study of Observational Research and Classroom Learning Evaluation (ORACLE) conducted by Galton & his colleagues (1980) in 58 classrooms and 19 primary schools near the end of the 1970s; the second was Mortimore, Sammons, Stoll, Lewis, and Ecob (1988) School Matters project; the third, published in the same year, was Tizard, Blatchford, Burke, Farquhar, and Plewis's (1988) Young Children at Schools in the Inner City which followed 343 children through three school years; the fourth was Pollard, Broadfoot, Croll, Osborn, and Abbott's (1994) PACE (Primary, Assessment, Curriculum and Experience) study investigating the changes that the 1988 Education Reform Act brought to schools, teachers, young learners and their perceptions of the new set of academic assessment, SATs; the fifth was the Effective Teachers of Numeracy (ETN) project carried out by Askew and colleagues (1997) from King's College London with a sample of 90 teachers and over 2,000 pupils; the sixth and most recent one was the Mathematics Enhancement Project Primary (MEPP) conducted by Muijs and Reynolds (2003) in 36 primary schools in England and Wales. Although overall TER studies carried out in the UK were much fewer than those done in the US, most strands of evidence further supported what had been proved to be effective teacher behaviours in the American literature.

In the remaining part of this section, the research designs and methodologies that the five large-scale studies utilised will be reviewed so as to shed light on the research design and methodology of this study; the findings of effective teacher behaviours that emerged in and considerably overlapped with each other among the five studies and other TER studies in the UK will then be analysed, integrated and linked with one another where possible. At the end of this section, research gaps will be summarised and presented in terms of methodologies and findings, and indications for this study will be concluded.

The research designs and methods utilised

The ORACLE study

As the first large-scale longitudinal observational study on teaching effectiveness in primary classrooms in the UK, the ORACLE study (Galton & Simon, 1980; Galton, Simo, & Croll, 1980) was conducted in 58 primary classrooms, across three local authorities, over three years. Pupil learning outcomes were tested against standardised testing papers for each school year. A huge amount of information was collected through 47,000 observations of 58 teachers and 84,000 observations of 489 pupils.

The discussion on the factors in relation to teaching effectiveness was mainly spread around teacher-pupil interactions, such as teacher questioning, teacher statements, audience and content of interactions, pupil time on task, teachers' attention to individual pupils and grouping strategies. The study found that the interactive whole-class teaching correlated positively with pupil learning outcomes/gains; that on average pupils in England spent two-thirds of classroom time working on their own; that of all interactions of teachers with pupils, 71.6% were with individuals; that on average teachers only spent 12% of lesson time interacting with the whole class.

The Junior School project

In their study, Mortimore et al. (1988) put their research boundaries around schools, making the educational effectiveness at both school and class levels their research focus. The core research questions fell on whether there were differences of effectiveness between schools as well as between classes and, if so, what factors contributed to such differences. Thus, this study examined both school effects and teacher effects on pupil learning progress.

The study followed 2,000 pupils (aged 7 to 11) from 50 schools through four school years. Though not overtly claiming to do so, the researchers seemed to have employed parts of the descriptive-correlational-experimental loop that the American researchers, Rosenshine and Furst (1973), recommended in their work. First of all, they entered the schools and classes and described what the situations actually were, with initial variables recorded, then built connections between variables and pupil learning gains over time and finally identified the effective factors that significantly contributed to performance progress.

Overall, the effective factors that the researchers found at the end of their study were located at two levels: the school level and the classroom level. Each level was then subdivided into two categories: '*given*' variables and *policy* variables. Thus, the effective factors were finally clustered into four domains: the school-level – 'given' variables, the school-level – policy variables, the class-level – 'given' variables, and the class-level – policy variables. The twelve effective factors at both school levels and classroom (teacher) levels were mutually supported and interwoven with each other. Teacher-level effective factors included: 'structured sessions', 'intellectually challenging teaching', 'work-centred environment', 'limited focus within sessions', 'maximum communication' and 'record keeping' (Mortimore et al., 1988, p. 250).

The Young Children at Schools in the Inner City Study

In order to investigate factors in relation to pupil learning outcomes and gains at early school years (Reception to Year 2), Tizard et al. (1988) conducted a three-year project in 33 infant schools of the Inner London Education Authority (ILEA). Four types of factors were focused on, including (1) background factors about children and their families, (2) parental involvement in home-based learning and home-school contact, (3) school and teacher related factors and (4) children's perceptions of their learning. Data were collected from pupils, their parents, teachers and headteachers. Schools were selected based on the criteria that each school should have pupils from multiple ethnic groups including children of Afro-Caribbean origins and white British. Research methods included interviews, observations and standardised tests. Observations were done at the pupil level. Within each class, two girls and two boys (*one Black and one white, each gender*) were chosen as target pupils to be observed for one school day per year.

The study found curriculum coverage, teacher expectations and pleasure to teach had positive impacts on pupil learning gains in mathematics, but there were no connections between pupil progress and home variables. Moreover, it was also found that curriculum coverage and teacher expectations were independently correlated with pupil learning gains to which teacher pleasure to teach was not independently related. It was concluded that factors at the school and classroom levels explained much of the variation of learning progress among pupils; that home and parental factors counted much less influence on pupil learning gains.

This study, again, proved the considerable effects of teaching on children's learning and also shed light on the impacts of ethnicity and gender characteristics on children's learning. Nonetheless, the highly selective nature of its sampling method limited the generalisability and reliability of its findings.

The Primary, Assessment, Curriculum and Experience (PACE) project

Following the implementation of the 1988 Education Reform Act, which aimed at raising pupil attainment standards, Pollard et al. (1994) conducted the PACE project. Through this project, the researchers investigated the impact of this educational policy on teachers' perspectives and their teaching practices and the subsequently indirect impact from teachers on the learning performance of Key Stage 1 pupils aged 5–7. Given the focus of this study was on the change that the newly implemented curriculum and the assessment, SATs, made to teaching and learning in schools, no sufficient evidence was found which could demonstrate the correlation between teacher behaviours and pupil achievement as normally seen in the process-product research on teaching effectiveness, except for the evidence on positive correlation between whole-class interaction and pupil task engagement.

The research methods, however, were rich in detail and worthy of notice and review. The data collection methods included two rounds of both interviews and questionnaires to teachers, and two rounds of classroom and assessment observations. The main sample of the study was 48 schools chosen with a balance between various social-economic factors, urban/rural areas and geographical locations in different regions of England. The main sample provided data for advance questionnaires and the first round of interviews. The sub-sample consisted of nine of the 48 schools from the main sample and contributed to the data for both systematic classroom and assessment observations.

The Effective Teachers of Numeracy (ETN) project

With data collected from 90 teachers and 2,000 pupils in 11 English primary schools deemed effective in maths teaching, Askew et al. (1997) sought to identify factors relating to effective maths teaching and possible ways for spreading such positive effects to wider contexts. Through the assessment of learning gains over two consecutive English school terms – Autumn and Spring terms, the study identified highly effective teachers. Data included a questionnaire with 90 teachers, lesson observations (84 lessons) in 33 teachers' classrooms, interviews with six headteachers, 54 interviews with 18 teachers and 30 validation interviews with 15 teachers.

The study had a particular focus on the relationships between teacher beliefs as well as teacher PCK and their actual teaching in the maths class. As a result, three orientations of maths teaching were identified: connectionist orientation, transformation orientation and discovery orientation. A major finding was that highly effective teachers tended to be connectionists who were able to interconnect different parts of mathematical knowledge and link the current status of children's learning and understanding to their subsequent teaching plans and practices.

Since teacher participants were from schools effective in maths teaching, the sample was not representative of the whole teacher population in England. Nevertheless, the study and its findings presented a picture of effective maths teaching in English schools in the mid-1990s and shed light on issues as to how and why certain kinds of maths teaching might have worked.

The ETN project was also one of the few studies that employed approaches from both the maths education research community (one that traditionally heavily favours the internal qualitative factors) and the teaching effectiveness research community (one that traditionally focuses more on external quantitative factors). The study marked a meaningful step forward for both fields by focusing on both internal and external factors.

Mathematics Enhancement Project Primary (MEPP)

Mathematics Enhancement Project Primary (MEPP, 1998–2001) aimed to transform the traditional British way of mathematics teaching in primary schools from teaching with a high proportion of individual work towards whole-class interactive teaching (Muijs & Reynolds, 2003). It measured the effectiveness of mathematics teaching in England and Wales with data collected through class observations (138 teachers) and standardised tests (4,813 pupils) in 36 primary schools. The *Mathematics Enhancing Classroom Observation Recording System* was utilised to evaluate maths lessons with 57 5-point teacher behavioural scales. Pre- and post-tests were done at the beginning and end of each academic year to provide accurate measurement of pupil learning outcomes and gains. The analysis focused on the relationship between teacher behaviours and pupil learning outcomes/gains. Multilevel modelling was applied to analyse variables at pupil, class and school levels, connecting these variances with both pupils' test scores at points in time and their learning gains over time. The class-level variables explained 24% of pupil achievement differences and over 50% of differences in learning gains over two years.

Unlike early TER researchers in the US, TER researchers in the UK rarely had the opportunity to carry out massive experiments in the class. Findings were almost all based on what was going on in the class at the time. One typical feature of the British class was the genuine interest in the individual child. There were repetitively identified effective teacher behaviours as well, but the amount was rather smaller than the early case of the US. Next, we will look at the *individuality* and *effective behaviours* emerging in the British literature.

The universal interest in the individual child

Across studies, the most commonly seen element of British teaching was teachers' strong interest in the individuals. The majority of lesson time was spent on interactions between the teacher and individuals or on pupils' independent seatwork.

The ORACLE study (Galton & Simon, 1980; Galton et al., 1980) found that pupils were left working on their own in about two-thirds of the class time and that teachers spent 71.6% of their limited interaction time interacting with individuals about routine things, most of which were to get children involved in their tasks. A similar finding from the Junior School Project (Mortimore et al., 1988) was that, on average, teachers spent 67% of lesson time in interacting with individual pupils, leaving just 23% of lesson time for whole-class interaction. In mathematics lessons, 13% of lesson time was spent on teacher questioning, whereas just a quarter of the questioning time, i.e. only about 3% of the lesson time, was applied at the whole-class level. The way in which university tutors interacted with individual students through personal tutorials was found to be extremely popular in the primary classrooms where around 29 other pupils could not be simultaneously engaged into the one-to-one communication with their teacher. Such individualised interaction was time- and energy-exhausting for the teacher and in fact unfair for all individuals.

The individualised teaching featured in the British classroom coincided with the progressive teaching approach recommended by the Plowden Report (Central Advisory Council for Education, 1967) which advocated child-centred education and individualised classrooms and followed Piaget's (1962) theory on the development of children's cognition. Nonetheless, as the researchers (Galton et al., 1980) argued, the individualisation recommended by the Plowden Report was not realistic for the classroom where around 30 children's needs were waiting to be met and teachers could hardly have sufficient time and energy to cope with each pupil individually. This argument was supported by the MEPP (Muijs & Reynolds, 2003) findings in that pupil learning outcomes and gains were both positively correlated with lesson time spent on whole-class interactive teaching.

Effective teacher behaviours identified

In the British TER literature, the most identified effective teacher behaviours included but were not limited to the following six: (1) applying most of the lesson time to whole-class interaction, (2) strong knowledge of subject matter, pedagogical content and learners, (3) pupils having more time on task, (4) optimising seat arrangements in the class, (5) high teacher expectations and (6) limited focus per lesson and sufficient curricular coverage over time.

Applying most of the lesson time to whole-class interaction

Among the factors that British research has identified as effective, whole-class interaction is the commonest. In the ORACLE study (Galton et al., 1980, Ch. 6), teachers were categorised according to their teaching styles as four main types: (1) individual monitors, (2) whole-class inquirers, (3) group instructors and (4) style changers. The last type was further classified as three sub-types: (4a) infrequent changers, (4b) rotating changers and (4c) habitual changers. The whole-class inquirers were found to only be focusing on the subject content they planned to teach, mostly teaching to the whole class, and therefore maximising their interactions with everyone. Follow-up analysis of the study (Croll, 1996) found a positive correlation between whole-class interactions and pupil learning outcomes. This was amongst findings of the MEPP study (Muijs & Reynolds, 2003) in that pupil learning outcomes and gains were both positively correlated with the amount of time for active whole-class teaching in lessons. The correlate was also evidence in other studies in the UK (Pollard et al., 1994) and the US (Evertson et al., 1980; Good, Grouws, & Ebmeier, 1983).

Strong knowledge of subject matter, pedagogical content and learners

The ETN project found that highly effective teachers had deeper understandings of mathematics, pedagogic content and pupils. They did not necessarily demonstrate a large amount of mathematical knowledge – "more is not necessarily better", but what they did have were deeper understandings of the connectedness of the subject knowledge (Askew et al., 1997, p. 98). They utilised assessment to systematically diagnose children's knowledge and then design their next-step teaching accordingly, which helped improve learning. Less effective teachers tended to merely care about whether the results were correct or wrong, and when children did not get the knowledge, they simply re-taught the stuff and asked for more practice.

Pupils having more time on task

Evidence from the ORACLE study showed a positive correlation between whole-class interaction and time on task (Croll, 1996). Similarly, in the *One in Five* study, which focused on the correlations between interaction types and academic engagement, Croll and Moses (1985) collected and analysed data from 32 Year-4 classrooms and found a considerable positive correlation between whole-class interaction and time on task and a moderately negative correlation between individual interaction and time on task. In the PACE study, Pollard et al. (1994, p. 182) found that, "classrooms where high levels of class interaction were recorded are also likely to have been coded with high levels of pupil task engagement."

Optimising seat arrangement in the class

Amongst several ways of seat arrangements, the horseshoe seating was found more effective, as it not only supported all the three types of teacher initiated interactions with individuals, groups and the whole class but also proved an easy way for the teacher to switch and maintain a balance between various types of interactions (Galton, Hargreaves, Comber, Wall, & Pell, 1999; MacNamara & Waugh, 1993).

High teacher expectations

Teacher expectations have long been identified as an effective factor that positively correlates with pupil learning outcomes/gains. In the Young Children at School in the Inner City Study, Tizard et al. (1988) found a statistically significant difference between pupils receiving high and low expectations from their teachers. A correlation between lower expectation and decreased pupil academic performance was also discovered in the study by Mortimore et al. (1988), as signals of lower expectations were transmitted to pupils through teachers' special attention based on pupils' SES, race and gender.

Limited focus per lesson and sufficient curricular coverage over time

More effective teachers tend to limit their focus to one or two curricular areas in a lesson yet cover more curricular content over time. Making comparisons across schools, Mortimore et al. (1988) found that more effective teachers oriented a lesson at no more than two curricular areas. In Tizard et al.'s (1988) study, a large curriculum coverage over time was found, along with higher academic expectations from teachers, posing positive effects upon the improvement of pupil performance. The researchers also found that teacher expectations tended to be transmitted to pupils through the curriculum content that the teacher assigned them to work on at class and/or at home.

Conclusion

The review of early British research on teaching effectiveness shows what methodologies and designs British researchers had generally employed and how well and what conclusions they had explicitly reached. The methodology commonly used was building connections between observed teacher behaviours and pupil learning outcomes/gains; the effective factors identified tended to work collaboratively rather than separately. During the reviewing process, the gaps in those two domains also naturally emerged. The methodologies and research designs of British TER studies – particularly those of the six major studies – formulated a picture with a wide scope. Building on often large-sized samples, TER studies in the UK inherited quantitative characteristics from TER in the US and continued identifying key teacher factors that had considerable impacts on pupils' learning. Those studies added extra value to the body of knowledge on what works and what does not in classrooms and schools. Nonetheless, there is still a lack of detailed explanations beneath the quantitative results. There are certain effective factors commonly identified in both the UK and the US. However, there are also many other factors that were less frequently identified and discussed by British researchers but were either frequently noticed by TER researchers from other countries or recommended by educational theorists and researchers utilising methods other than the product-process approach. The next section thus focuses on TER in other countries.

Teaching effectiveness research in other nations

International studies of pupil performance show the disparity of educational outcomes between educational systems across different countries. In parallel with the international investigations on teaching and schooling among countries and systems, there has been an on-going trend within a number of countries to examine factors that decide the effectiveness of their national education.

The aforementioned Hattie's (2003) review on half a million studies comes up with five dimensions and 16 attributes of effective factors that expert teachers generally have. The review suggests that effective teachers are able to represent the subjects optimally, create an appropriate environment for best learning to happen, diagnose and tackle learning difficulties, enthuse about pupils and teaching and learning and engage children towards effective learning.

Involving 50 Australian schools, Ingvarson, Beavis, Bishop, Peck, and Elsworth study (2004) focused on the correlations between teacher knowledge and pupil learning outcomes. Data were collected through two tests, questionnaires with pupils, teachers, headteachers and maths departmental leaders and case studies. The study found that teachers' educational background and the professional quality of the maths department were strong predictors of learning gains and that teaching methods predicted pupil perceptions of maths learning.

Starting from the theories of learning, Munro (1999) followed 32 Melbourne teachers through the process of a professional programme, called Facilitating Effective Learning and Teaching, and identified the changes of teaching behaviours that coincided with the learning theories embedded in this programme. The aim of the programme was to promote teachers' reflection on pupils' learning in class under a framework containing nine effective components so that teachers could improve their teaching accordingly. The nine components included (1) promoting effective learning through managing pupil attention,

(2) boosting pupil interests in subject learning, (3) offering rich options in a topic to meet learning needs at different levels, (4) allowing pupils to present their learning in various ways, (5) enhancing self-efficacy in learning, (6) "using learner-oriented behaviour management strategies" (*ibid.*, p. 162), (7) nurturing meta-learning skills, (8) nurturing collaboration amongst pupils in managing their own learning and (9) increasing pupils' awareness of their learning and progress. Pre- and post-measurement was done in terms of changes in three domains: changes in teachers' behaviours, changes in teacher knowledge and changes in pupil learning outcomes. The study found a positive correlation between improved teaching behaviours and learning gains.

In a Dutch study conducted by Luyten and de Jong (1998), teacher effects and school effects were examined in the subject of mathematics in parallel classes within and across 20 secondary schools. Through structured interviews and common tests in parallel classes, pupil achievement was compared at pupil, teacher (class) and school levels. Apparent differences were found within classes, while pupils from parallel classes performed at a similar level. Measured against a five-aspect model, which contained structured instruction, clarity, pupil enjoyment, teaching content and homework quantity per week, mathematics teachers were found significantly different from each other within schools in all five aspects, particularly in terms of structured instruction and homework quantity per week.

Some other Dutch researchers are interested in the impact of teacher-pupil relationship on academic and/or affective outcomes. The study by den Brok, Brekelmans, Wubbels (2004) looked into the correlations between teachers' interpersonal behaviours as perceived by pupils and pupils' learning outcomes. Findings were based on data derived from 45 physics teachers, 32 English teachers and their Year-3 pupils. Two pupil-perceived teacher behaviours, *influence* and *proximity*, were found responsible for more than half of the differences in learning outcomes. Overall, physics teachers' interpersonal behaviours had a stronger impact on affective outcomes but a weaker impact on academic outcomes than did English teachers'. This is consistent with findings from other studies in that teacher effects vary between different domains of outcomes and between various subjects (Luyten, 1998; Reynolds, 1995). Interested in the same types of teacher behaviours, Opdenakker, Maulana & den Brok (2012) found a decrease of teacher *proximity* and *influence* over time, the impact of class type and teacher gender on *proximity*, an increase of pupil-controlled academic motivation and a decrease of autonomous motivation.

In Germany, Einsiedler and Treinies (1997) conducted a study which combined both experimental and observational data. Twenty-one Grade 4 classes were randomly and averagely divided into three groups of classes which were taught by teachers using three different representational methods, i.e. hierarchical knowledge structuring, network-like knowledge structuring and traditional knowledge structuring (in the control group). In the experimental section of this study, researchers found that pupils above average benefited more in the treated groups, but pupils below average progressed more in the controlled group. However, overall, it was found that class effects were stronger than the effects of teaching methods, i.e. the representational methods. Utilising Flanders' method (1970), trained observers recorded the frequencies of various types of verbal cognitive interactions between teachers and pupils in the 21 classes. Consequently, the observational data revealed more directions in the two treated groups than in the controlled group. In addition, teachers in treated groups tended to structure lessons with graphic representations, while teachers in the controlled group were more likely to apply cognitive structuring. Overall, the researchers stressed the importance of class contexts to the effects of school education and the need to take into account the role of teachers in formulating the contexts of classes.

In a project on stakeholders' perceptions of education quality in China, Peng et al. (2014) conducted individual and focus group interviews with over 90 stakeholders, including headteachers, teachers and high school students from eight secondary schools across three LEAs in China. Partial findings of the study indicated that Chinese teachers were expected to play multi-functional roles, that they experienced dilemmas in meeting the demands of the curriculum reform, that the formulation of a positive learning environment in a school was expected to be a collaboration between teachers and pupils, that within Chinese schools there existed professional learning communities which aimed at improving teaching through collaborations between teachers, their colleagues and/or external experts from teacher education institutions. The study also found a disparity of educational development between western rural areas and other parts of China and the dilemma that stakeholders in the west of China faced as they strove to improve teaching and learning in their schools.

Another noteworthy TER study was conducted by Teddlie and Liu (2008) in a relatively poor province, Jilin, in Northeast China. The sample included teachers and pupils from eight urban and four rural schools. Collecting data through observations of 300 lessons in five grades (i.e. year levels) across 12 schools, the study found that effective teaching factors identified in the West actually worked in these Chinese schools, that more effective schools had teachers demonstrating more effective teaching practice than less effective schools, and that urban schools tended to have more effective teachers than rural schools. The study identified effective teaching factors, such as nurturing positive learning environment, maximising the proportion of lesson time spent teaching, managing pupil behaviours and interactive time-on-task. A set of teaching factors, including whole class interactive teaching, was believed to have guaranteed the high average of interactive time-on-task (M = 81%) – a strikingly much higher score than the combined American score (44%) in the literature. However, whether the performance differences between countries (such as China and the US) were attributed to teaching differences could only be answered by *cross-national* TER studies.

In this section, we have looked at TER studies carried out in Australia, the Netherlands, Germany and China. Due to word limit, many other studies from Belgium (e.g. Boonen, Van Damme, & Onghena, 2013), Cyprus (Kyriakides,

Creemers, & Antoniou, 2009), the Netherlands (Driessen & Sleegers, 2000; Meijnen, Lagerweij, & Jong, 2003) and other countries (Konstantopoulos & Sun, 2013; Palardy & Rumberger, 2008; Teodorović, 2011; van de Grift, 2013) have not been reviewed here.

Chapter conclusion

This chapter reviews the evolution of educational effectiveness research, early research on teaching effectiveness in the US, early research on teaching effectiveness in the UK and other national studies on teaching effectiveness.

The field of EER has evolved vertically through over four decades and horizontally across locations – classrooms, schools and countries. Similar and different effective educational factors have emerged among EER studies, most of which were conducted within countries. The flourishing of EER in more countries offers a good opportunity for researchers in the field to carry out international studies to examine what factors work internationally and what may work better locally. In terms of paradigm orientation of the field, there is a severe lack of rich descriptions and explanations in the outputs of EER where quantitative approaches have been playing a dominant role and qualitative evidence has rarely been collected or equally appreciated. Decades of EER knowledge base consistently proves the crucial role of classroom-level factors, particularly teacher behaviours, in fostering the improvement of pupils' learning. In line with the international comparisons of educational outcomes at similar ages, there is a need for TER researchers to conduct cross-national studies on the effectiveness of teaching in order to generalise internationally valid and reliable predictors for pupil performance and promote the development of and universal applicability of educational science (Reynolds, Creemers, Stringfield, Teddlie, & Schaffer, 2002a).

The early research on teaching effectiveness in the US and the UK sees a slight shift from the process-product research towards not only the correlations between teacher behaviours and pupil learning outcomes and/or gains but also deeper inquiries into the reasons beneath the surface of various correlations. Later research on teaching effectiveness in other countries shows a diverse array of research foci and inquiry approaches. Mixed-methods gradually came into the research trend to complement the quantitative methods with great detail collected via qualitative methods. This reconciliation between two paradigms adds into the EER and its subfields down-to-earth properties and initiates the trend in which the EER and the educational improvement research teams join each other to make a greater contribution to educational theory, policy and practice.

In this study, in answering the research questions, we set off from the mixed-method paradigm and reached a new research paradigm, the integrated paradigm, which guided the analysis and synthesis of a variety of data. The following chapter gives the details about the paradigm, design and methods of the study.

4 Research paradigm, design and methods

Chapter overview

- Research philosophy
- Research design
- Structured lesson observations (M1)
- Questionnaires to teachers and pupils (M2)
- Standardised mathematics tests (M3)
- Unstructured lesson observations (M4)
- Interviews with teachers (M5)
- Focus groups with teachers (M6)
- Integrating findings across methods

The research gaps that this study intends to fill in and the three main research questions that the study aims to address emerged from the review of relevant literatures. In this chapter, we will focus on the technical part of the project to explain in detail:

- the research philosophy in which the study was embedded,
- the research design which defined and was defined by the study,
- specific data-collection and data-analysis methods applied,
- the way different strands of findings were integrated.

Research philosophy

Before we set out to find answers for the research questions, we reflected upon different streams of research philosophy in education. On the one hand, there was the polarisation between two classical paradigms – the positivism (favouring quantitative methods/data) and the interpretivism (endorsing qualitative methods/data); on the other, there was the gradually rising mixed-methods approach. We have started from the third paradigm, but the data analysis process saw the emergence of a new paradigm which we now call the integrated paradigm. Looking back, we realise that none of the three paradigms seem competent in representing the teaching and learning reality we propose to capture. The positivism paradigm might be scientific yet too abstract and inhuman, whilst the interpretivism is capable of collecting vivid stories but cannot generate rigorous and widely applicable solutions. The third paradigm – mixed-methods – is suspicious of seeking a compromise between the two old paradigms.

Amidst data analysis, we increasingly felt a necessity to push the boundary further to invite this whole new paradigm – the integrated paradigm – to guide the synthesis of the results and findings of the study. In the wake of the birth of the new paradigm, the traditional paradigm war ceased, and different strands of data and findings started to make sense, as different parts of evidence rose and joined each other to represent the whole reality. In the integrated paradigm, we zoomed in to capture details of the reality with numbers (QUAN) and voices (QUAL) and zoomed out to view the whole picture and how the picture was formed with all these details mutually interconnected. The philosophy underpins the research questions, methods, theoretical framework, processes and findings.

Research design

As presented in Chapter 1, the aims of the study are, on the one hand, to test the correlations between teaching behaviours and pupil learning outcomes across England and China, and on the other, to listen to and interpret multiple voices regarding the quality and effectiveness of mathematics teaching within and across these two countries. In doing so, QUAN methods were mainly applied to answer Q1, whereas QUAL methods were specifically chosen for Q2 and for the analysing process towards a meaningful conclusion of Q3.

Given that the research aim is to investigate the effectiveness of mathematics *teaching*, the classroom (teacher) level factors are put at the core of the research focus. Research methods applied include structured observations of maths lessons (M1), questionnaires to teachers and pupils (M2), standardised mathematics tests (M3) for pupils at two points in time, unstructured lesson observations (M4), video-stimulated interviews with teachers (M5) and video-stimulated focus groups with teachers (M6).

Based on the research questions, methods applied and interconnections between research questions and methods and between results and findings of the project, a theoretical framework is constructed (Figure 4.1). The framework links research questions to research methods, with the left hand side of the framework standing for the first question, the right hand side the second one and, in between, question three.

The theoretical framework is more than just a framework. It is also a research map that guides every step of the study. Technically, it not only structured the data collection and data analysis processes, but also illustrated the interconnections between various research methods and different components of data. Philosophically, it is a representation of reality where numbers as well as voices,

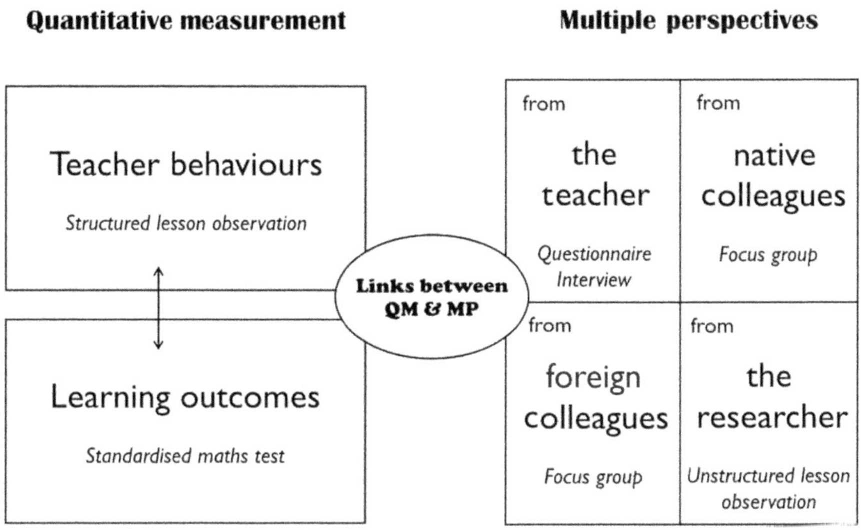

Figure 4.1 The EMT theoretical framework

quantity as well as quality and external behaviours as well as internal perceptions inter-exist. The framework plays multiple roles in the study, in that it explains the philosophy on which the study stands, links the research questions to the research methods and explicitly guides the preparation, implementation, analyses of and subsequent conclusions of the project.

With this research map handy, the next thing to do is to find participants. Given the limitation of a PhD study in terms of resources, a combination of convenience and stratified sampling was applied. Convenience samples are usually obtained for easy access. Conventionally, this approach is not probability sampling. However, the combination of it with probability sampling increases the statistical strength of the combined approach. Stratified sampling is a way of probability sampling in which the population is divided into "homogenous groups, each group containing subjects with similar characteristics" (Cohen, Manion, & Morrison, 2007, p. 111). With "a useful blend of randomization and categorization" (*ibid.*, p. 112), this sampling method allows the researcher to collect data that can not only be analysed quantitatively, but also be interpreted qualitatively. Focusing on a specific stratum also helps limit the number of researchers, amount of time and other resources that a study may need.

Then a PhD student (the first author) from China and a professor/supervisor (the second author) based in the University of Southampton, England, we are both interested in international comparisons and what works in the teaching of mathematics cross-nationally. England and China naturally come to the fore as potential contexts for the study. In order to compare like with like, it is considered reasonable to compare cities that are at the medium-to-high level in socio-economic terms in both countries. To limit the data to a manageable size, it is proposed that, in each country, data will be collected in one city representing the average of the medium-to-high economic stratum of cities in its home country. A focus on the medium-to-high level cities is due to the fact that China is still a developing country where the gap between rural and urban regions in terms of educational resources can be considerable with the latter generally better equipped than the former, though the country has policies to support less developed regions (Peng et al., 2014).

The first author is from *Nanjing* and is studying in *Southampton*. It is thus proposed to find out whether the two familiar and of course convenient cities match the criteria. Subsequently, a series of national and municipal data searches has been carried out, which confirms the comparability of these two nations and cities in socio-economic terms. Key information regarding the comparability and representativeness as well as participating schools is presented in Chapter 1, which also works as a bird's-eye view of the research contexts.

At the heart of the stratified sampling method lie the criteria that define where and how participants are recruited. The EMT project's focus is on the population of teachers whose pupils (aged 9–10) are at the average level in medium-to-high level cities in each country. In this study, by *average level*, it means that two criteria are considered: (1) pupils' academic performance and (2) their socio-economic status (SES).

Systematic sampling procedures on both sites resulted in the participation of three schools in Southampton and four schools from Nanjing. All maths teachers (N = 19) in Grade 4/Year 5 from each participating school took part in the study. For the 9 classes (number of pupils in all events = 343) from four schools in Nanjing, two of them could not make the first test, because of timing difficulties, so seven sat the first test, and all nine classes (number of pupils in test two = 326) took the second test. Table 4.1 shows the final sample and the responding rates of each data-collection event.

The following sections give detailed descriptions of research methods that were applied.

Table 4.1 The final EMT sample and the response rates

	M1 & M4	*M2*	*M3*	*M5*	*M6*
EN pupils (n=236)	236 (100%)	231 (97.9%)	T1: 231 (97.9%) T2: 236 (100%)	–	–
CN pupils (n=343)	343 (100%)	343 (100%)	T1: 250 (72.9%) T2: 326 (95%)	–	–
EN teachers (n=10)	10 (100%)	10 (100%)	–	10 (100%)	10 (100%)
CN teachers (n=9)	9 (100%)	9 (100%)	–	9 (100%)	9 (100%)

Structured lesson observations (M1)

In each of the 19 classrooms, a maths lesson was observed. Teachers were well aware that what we really wanted to see was a lesson that represented what they had been doing in the maths class every day. Three video cameras were utilised to record lessons (n=19) in both countries. The researcher generally arrived at schools half an hour earlier to get familiar with the teacher, pupils and classroom.

Cameras were placed at suitable places where ideal pictures could be captured and where pupils were less likely to walk around. In order to shoot the lesson with the teacher's, pupils' and researchers' perspectives all considered, three cameras were set up in every classroom. These included two smaller handycams (Cam #1 and Cam #2) and a larger-size professional camcorder (Cam #3). Each camcorder was attached to a tripod. Cam #1 and Cam #2 were set up pointing to two randomly chosen groups of pupils to capture their activities throughout a lesson. In most cases, they were placed at each side of the front board where the teacher generally gave demonstrations to the whole class. Once the two small handycams were set up and switched on before the lesson started, there would be no other operations on them throughout the lesson. The researcher was then free to concentrate on the operation of Cam #3. Cam #3 was placed near the wall *opposite* to the front board to offer a wide-angle view of the whole class. Through Cam #3, the researcher was trying to view the lesson from both the points of view of pupils and an outsider – the researcher. Mostly, it kept following the teacher and occasionally zooming in to see details, for example, about the teacher's marking on a pupil's work.

Lesson videos were analysed in both structured and unstructured ways. For the structured analysis, lessons were measured with two existing observational systems developed by two international research teams who have conducted educational effectiveness research cross-nationally. One is called Opportunity to Learn (OTL) measuring the quantity of teaching, and the other the International System for Teacher Observation and Feedback (ISTOF) measuring the quality of teaching.

OTL was adopted from the International School Effectiveness Research Project (ISERP) (Reynolds et al., 2002a) – an SER study across nine countries and four continents (America, Asia, Australia and Europe). It has effective school factors and effective teacher factors in different contexts as the first of its three research questions. The ISERP research team designed OTL as one of the nine optional instruments and protocols for research at pupil and teacher levels. OTL has a particular focus on percentages of time allocation to five different activities (in the left column) and percentages of pupil time on task (in the right column). The five OTL activity categories are *whole class interaction* (OTL-1), *whole class lecture* (OTL-2), *individual/group work* (OTL-3), *classroom management* (OTL-4) and *partial class interaction* (OTL-5) (replacing testing/assessment).[1] Plus *pupil time on task* (OTL-6), OTL has a total of six measures all in the form of percentages.

ISTOF was designed, applied and validated in two empirical studies across over 20 countries by Teddlie, Creemers, Kyriakides, Muijs, and Fen (2006). The original ISTOF protocol contains seven domains. Within each domain, several subdomains (indicators) were listed, which in turn were subdivided into various teacher behaviours. Overall, there are 45 teacher behaviours measured on a 5-point Likert scale, ranging from "1 = strongly disagree" to "5 = strongly agree". In this study, one domain named *Differentiation and inclusion* (4 items) and one differentiation-relevant item from the *Instructional skills* domain were taken out. The decision was made not simply because differentiation was unimportant but because all Chinese primary schools were mixed settings within and across classrooms. Chinese teachers are dealing with the ability differences among pupils in a different way from what the five items have described, hence the decision. The revised ISTOF contains 6 domains: *Assessment and Evaluation* (ISTOF 1), *Clarity of Instruction* (ISTOF 2), *Instructional Skills* (ISTOF 3), *Promoting Active Learning and Developing Metacognitive Skills* (ISTOF 4), *Classroom Climate* (ISTOF 5) and *Classroom Management* (ISTOF 6). In total, there were 40 behavioural scales.

Two observational systems supplemented with each other, in that they evaluated lessons at both micro (OTL) and macro (ISTOF) levels and examined both quantity (OTL) and quality (ISTOF) of teaching. As lessons were all video-recorded, the traditional *on-site* ticking and scaling on a structured schedule were postponed for *off-site* analyses on lesson videos. Before formal analyses, the two authors carried out trial observations on four lessons, two from China and another two from England, reaching the inter-rater reliability (Cohen's Kappa) at 0.78, 0.81, 0.84 and 0.88, respectively. An average score of 0.83 (> 0.7) suggested that the reliability was achieved. The formal process involved ratings and data input. Raw results from both systems were put into Excel and further analysed in SPSS. Teacher variables consisted of the six OTL measures and six ISTOF measures.

Questionnaires to teachers and pupils (M2)

We gathered information regarding teacher beliefs, teacher self-efficacy, teachers' self-evaluation of their teaching and school climate with revised version of TALIS 2008 questionnaire (OECD, 2009a, 2009b). Pupil demographic information and their perceptions of learning and schooling were also surveyed with the background information items (#68-#80) from the pupil questionnaire of the MET project (Bill & Melinda Gates Foundation, 2012).

Both questionnaires were checked and revised where necessary to ensure they were in British English and made sense to the English context. To get a valid Chinese version, the two instruments were each translated into Chinese and then translated back to English so that the Chinese instruments carried exactly the intended information. After the translation and back-translation process, a Chinese pupil and a Chinese teacher were asked to proofread the translated pupil questionnaire and teacher questionnaire, respectively.

Standardised mathematics tests (M3)

At the heart of TER sits the correlation between teacher behaviours and pupil achievement/attainment. This had rarely been conducted comparatively in the proposed contexts, England and China. For this study, two standardised mathematics tests were proposed to examine mathematics learning outcomes at points in time and learning gains over 10 school weeks with the same set of test items. We consider the released TIMSS 2003 items (4th-grade maths) (IEA, 2007b) as appropriate for the EMT project since it was designed for children at the same age as the EMT cohort (9- to 10-year-olds) and it had been validated and applied internationally.

To compile a test paper for the project, we took the following aspects into consideration: (1) the percentages of various content domains and cognitive domains defined in the TIMSS 2003 Assessment Framework (Mullis et al., 2003), (2) a balanced coverage of main topics within each content domain, (3) the percentages of various item formats, (4) the timing of the tests and (5) social and cultural adaptability of the EMT test paper. The finalised paper contained 40 items to be completed in 40 minutes. Despite every confidence we had regarding TIMSS 2003 items, the test results suggested later on that the items were unable to measure learning gains in both countries, since they were too easy for the Chinese pupils whose progress had been significantly stopped by a glass ceiling. Test results on two occasions were thus separately correlated with teaching measured with OTL and ISTOF.

The two paper-based tests were marked and double-checked by the researcher following the TIMSS 2003 marking guidance (IEA, 2007a). The score of every paper was the percentage of correct responses the pupil made. The correct rates of every content domain and cognitive domain were also calculated to generate performance patterns within and across countries. Pooling two countries' classroom-level data together, we calculated Pearson's r correlation coefficients with the quantity (OTL measures) and quality (ISTOF measures) of teaching as x and class-level mean scores in each test as y.

Unstructured lesson observations (M4)

Lesson data were also analysed in an unstructured way. Free from predefined categories and scales, the purpose was to gain deeper understandings of the *flow* (i.e. process) and *characteristics* of maths teaching in each country.

Interviews with teachers (M5)

In the EMT study, teachers not only carried out lessons for external evaluations on the effectiveness of their teaching but also gave their own comments on the observed lessons while watching the lesson video in a post-lesson interview. Interview data provided rich information about why they were teaching in the way they were. Lesson videos clarified the focus of interviewing and took teachers on a journey back in time to revisit their lessons. The advantage of video-stimulated interview is evident in a number of previous studies, such as Learner's Perspective Study (Clarke et al., 2006) and Preschool in Three Cultures (Tobin, Hsueh, & Karasawa, 2009; Tobin, Wu, & Davidson, 1989).

In order to guide the conversation between the teacher and the researcher towards the research focus, an interview framework was built containing questions relevant to the focus. The framework consisted of eight questions which were clustered around three concepts: (1) teacher beliefs (itvQ1-itvQ3), (2) teacher self-evaluations [itvQ4-itvQ6] and (3) international awareness and attitude to practice reform [itvQ7-itvQ8]. Questions were about:

- teacher's personal beliefs/descriptions of an effective mathematics lesson (itvQ1),
- teacher's strategies for organising and managing lessons (itvQ2),
- teacher's strategies for differentiation so as to meet various learning needs (itvQ3),
- teacher's teaching plan (itvQ4),
- teacher's reflection on how the plan went (itvQ5),
- teacher's comments on the lesson and its effectiveness while and after watching the lesson video (itvQ6),
- teacher's imagination of a mathematics lesson to the 9- to 10-year-olds in the other participating country (i.e. England or China) (itvQ7),
- teacher's possibility of changing their teaching beliefs and practices in the future (itvQ8).

The purpose of asking the first and second sets of questions was to understand the connections between teacher beliefs and teacher self-evaluations. The second question set also served as one of multiple perspectives on the observed lesson. The third question set facilitated individual teachers to foresee maths teaching home and abroad before actually seeing it in the focus group.

Though these questions were all predefined, the answers to them were completely open, hence making the interviews *semi-structured*. In order to ask questions when each lesson was still fresh in a teacher's mind, every interview was scheduled for the same day or the day after the observation.

At the analysis stage, interview data were organised by groups (nations) and issues (interview questions) as suggested by Cohen et al. (2007, pp. 467–468). International similarities and differences were discussed where necessary.

Focus groups with teachers (M6)

The purpose of conducting focus groups was to collect teachers' views on two maths lessons – one from England and the other from China. Within each country, teacher participants were divided into two groups. There were five teachers in each of EN-FG1 and EN-FG2 and initially six teachers in each of

CN-FG1 and CN-FG2. All data of three teachers in CN-FG2 were excluded from the data set, as their school could not participate in the tests.

Before focus groups, both lessons were translated into the corresponding foreign language – British English (the Chinese lesson) or Simplified Chinese (the English lesson) – to make the content understandable in either context. The process of each focus group followed the flowchart shown in Figure 4.2.

The interpretation of focus group data followed the similar approach to analysing interview data as suggested by Cohen et al. (2007, pp. 467–468). Data were organised by groups (nations) and issues (lessons).

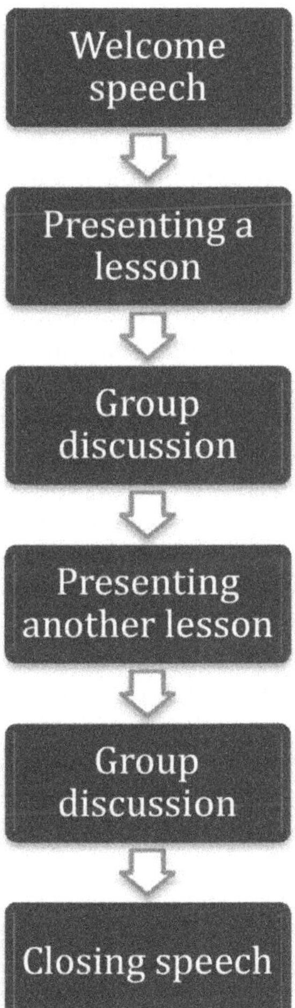

Figure 4.2 The flowchart for focus groups in the EMT project

Integrating findings across methods

Through the lens of each method, we saw a piece of maths teaching and learning reality. Initially, we focused on and presented every strand of evidence. Then, we went deeper to seek details/descriptions not only in 'voices' (QUAL) but also in 'numbers' (QUAN) and abstract patterns/frequencies, not only in numbers but also in voices. Finally, we stepped back and forward and back again and zoomed out and in and out again to connect the 'dots' in every sensible way to represent the whole reality in great detail. Apart from the integration of different types of data and findings, the study also integrates two lines of research on mathematics teaching – one focusing on factors within teachers' minds (thinking and knowledge) and the other on factors without (behaviours).

Chapter conclusion

This chapter clarified the philosophical position of the study, located research methods and research questions in the theoretical framework, and explained the methodological details of the study by linking research questions to research methods. The design of the study also benefits from the methodological lessons from previous studies in similar areas as reviewed in Chapters 2 and 3. Succeeding the methodological heritage, this study contributes to the methodological evolution of teacher effectiveness research in the subject of mathematics.

In this study, the application of multiple methods in addressing research questions brings together scientific evaluation of teaching and learning and multiple perspectives from various roles in and beyond the classroom. This in turn will contribute to future collaborations between educational effectiveness researchers and educational improvement researchers with their methods mutually supplemented in one study. The methodological emphasis in this study lies on the interconnections between multi-layered data and findings generated by these methods. The design of the study is purpose-driven, instead of paradigm-driven. Having said that, the study discovers a new paradigm – the integrated paradigm where all types of methods, data and results and findings interrelate at different layers throughout all stages of research to represent a holistic reality with both external and internal, objective and subjective, observable and latent details.

Results and findings are to be reported in the following chapters. We will start by introducing the participants, children and their teachers, who gave their information in the pupil questionnaire and teacher questionnaire.

Note

1 Partial class interaction was repeatedly discovered in almost all English classrooms during the initial lesson data analysis. Testing/assessment behaviours were always found netted in either whole class or partial class interactions in English lessons and with whole class interactions in Chinese lessons. It was thus decided that "partial class interaction" would replace "testing/assessment" as the fifth component of OTL to better describe the quantity of teaching.

5 Children and their maths teachers

Chapter overview

- Children as individuals
- Children as classes of pupils in schools
- Maths teachers – the demographic selves
- Maths teachers – the professional selves

This chapter introduces children and their teachers with the information derived from the questionnaires that they filled in. Children are introduced both as individuals with their own demographic features and as classes of pupils taught by teachers in schools. Likewise, teachers are profiled with both demographic and professional characteristics.

Overall, 579 children from England (236) and China (343) participated in the EMT study, along with their maths teachers. With a participation rate of 99.1% in the pupil questionnaire, we are able to seek an understanding of these children and their peers from the same social stratum, same country, in terms of backgrounds, maths learning and teaching as well as schooling. In a neutral stance, we attempt to reach and deepen the understanding through the "comparative" lens.

Children as individuals

In either country, the proportion of boys is slightly more than that of girls, but the difference is bigger in China than in England, in that the former has a girl-boy ratio of 43.8:56.2 and the latter 49.3:50.7.

There are more percentages of children coming from ethnic minority background in England (17.5%) than in China (3%). This suggests that English children are living in a more inclusive and diverse society.

The results of Teacher Questionnaire (item #36) suggest that Chinese children are more likely to have parents with a higher level of education than English children. According to teachers' estimates, no more than 20% of English children's parents have a Bachelor degree or higher which is the case of 20%

to 40% of parents of Chinese children in three classes, 40% to 60% in one class, and over 60% in four Chinese classes.

The majority of children have access to one or more computers at home. Nevertheless, there is a notable difference between the two countries' proportions of children who do not have a computer at all. This is the situation of about 8% of English children in comparison with 2.4% of Chinese children.

In either country, over 90% of children have a varying number of books in their bedrooms. However, the proportion of English children who do not have any books in their bedrooms tripled that of the corresponding Chinese children. Approximately 30% more Chinese children have 25 or more books in their bedrooms than the proportion of English children with this number of books. This suggests that Chinese children may have more opportunities to read and presumably a more supportive learning environment at home than their English peers.

English children are more likely to have siblings than Chinese children. Three-quarters of Chinese children are the only child at home, whereas this is the case for only 17.5% of English children. Over half of English children have one or two siblings. It is clear that Chinese children have a lack of experience of sisterhood or brotherhood at home in comparison with English children. However, it is not clear and of course beyond the research focus of the study how the only child experience might impact on Chinese children's study and life.

Though the majority of children from both countries have two or more adults at home, the questionnaire results also suggest that the proportion of English children (15.5%) coming from a single parent family is about five times bigger than the Chinese proportion (2.7%). Moreover, Chinese children (44%) are more likely to live with parents and grandparents (or other family members) under the same roof than English children (11.1%).

Whilst there are about one-third of children in each country holding a negative view regarding their previous performance in the subject, Chinese children are slightly more likely to think highly of their previous performance in mathematics than their peers in England. This suggests a higher self-esteem amongst kids in China than England.

To summarise, Chinese children were more likely to have no siblings than English children who usually had about two siblings or more. In both countries, those who had no computers at home only counted less than 8%. However, the English proportion of children in this circumstance tripled the Chinese figure. Moreover, 27% more children in the Chinese sample reported having more than 24 books in their bedrooms than the English proportion. In addition, according to teachers, Chinese parents seemed to have higher levels of education than English parents.

The overall SES data seem to suggest that Chinese children might come from a slightly more advantaged background than their peers in England. These pupils' background differences might have affected teacher behaviours as well as children's maths performance, but there is nothing available in worldwide literature on the determinants of educational outcomes that would suggest the relatively small differences between England and China in these areas would be able to explain the very large differences in educational outcomes, teacher behaviours, classroom factors and all the other "process" variables. To the researchers, it is clear that educational processes – not just background factors – are strongly implicated in the educational differences.

Children as classes of pupils in schools

It is observable in the classroom that Chinese children have a more disciplined environment for learning than do English children. Teacher questionnaire results, however, offer a complex picture. Teachers, in responding to the questionnaire item #39, claim a varying degree of the existence of noise across classrooms. Half of English teachers perceive much noise in the classroom and only one-eighth of Chinese teachers encounter the same problem. In fact, some English teachers do not consider noise a problem. The striking contrast is that one-eighth of Chinese teachers also indicate that they have to wait for a long time for pupils to quiet down before lessons start, whilst none of English teachers say they have to do so, which is, however, not observable in classrooms at all. It appears that English teachers are more optimistic about classroom disciplines than the Chinese, apart from their noisy classrooms for which some English teachers have applauded with clear-cut words written in the questionnaire, for instance, "Noise is good!"

According to the teachers' questionnaire responses, English classes have a much wider spread of pupil abilities than Chinese classes. *All* Chinese children are regarded as *about average* or *higher than average* in comparison with their peers at the same age both within and beyond their schools. The ratings of English children's abilities vary from *much lower than average* to *much higher than average*. This contrast of ability ranges between countries echoes with the international difference of the spread of performance in the two maths tests.

England is usually said to have a relatively smaller class size than China, which is evident in this sample. On average, the English classroom caters for 27 children ($SD = 3$) and the Chinese classroom 38 children ($SD = 5$).

Over 99% of children in each country express that they have a certain amount of homework. It is obvious that Chinese children receive homework more frequently than English children. Almost all Chinese children report that they have homework five days a week, whereas the corresponding proportion for England is just slightly over a quarter. Half of English children have homework one day per week, with the remainder evenly (5% each) falling into the frequencies of 2 days per week, 3 days per week and 4 days per week. Moreover, Chinese children are more likely to complete homework than their English peers. As suggested by the Pupil Questionnaire, 83% of Chinese children complete all or all plus extra homework, and so do 73% of English children. Overall, Chinese pupils have more homework on almost every school day and are also more likely to complete their homework than English pupils.

Regarding their self-esteem, pupils were asked, "When you were younger, what kind of marks (or grades) did you usually get in school?" Fifty-one percent of Chinese children and 42% of English children rated their previous performance as *high* or *very high*. It seems that Chinese pupils are slightly more confident than their English buddies.

The majority of all pupils have been in their current classes for over two months, but a notable fact is that the proportion of the English children joining the class within two months is double the corresponding Chinese proportion.

As aforementioned, 10 English maths teachers and nine Chinese maths teachers took part in the study. This section forms a *sketch* of the teacher participants from both countries with *lines* of demographic information and information about their professional selves.

Maths teachers – the demographic selves

Of teacher participants from either country, two-thirds are females, and one-third males; nine English and eight Chinese teachers are in their twenties or thirties, one English teacher is under 25, and one Chinese in his forties.

Maths teachers – the professional selves

Apart from one English teacher with a Master's degree, all other 18 teachers have a Bachelor's degree. Holistically, according to "the professional age", England seems to have a younger teaching team. Seven English and three Chinese teachers have one to five years' experience; two teachers from each country have 6–10 years' experience; one from each country has 11–15 years' experience; three Chinese teachers have been teaching for 16–20 or more years. Though all teachers are full-time, the majority (9 of 10) of English teachers have permanent positions, but the majority (6 of 9) of Chinese teachers are on a fixed-term contract.

The comparison between the number of years into teaching and the number of years into teaching in the current school suggests that English teachers are more likely to stay in the same school where they started their career than Chinese teachers. Three Chinese teachers with the experience of 16–20 or 11–15 years' teaching have moved to the current school for 3 to 5 years, whereas only one English teacher with a total of 3–5 years' experience has recently moved to the current school.

Professional development

Over the past 18 months, English teachers have spent an average of approximately six days on professional development, and this figure for Chinese teachers is 45. In either country, teachers who take part in professional development activities outside regular work hours do not normally receive a salary supplement for doing so. Apart from one type of development activities – *research activities* – in which none of English teachers and four of nine Chinese teachers have taken part, a varying number of teachers from either country have spent time on *course/workshops*, *education conferences/seminars*, *qualification programmes*, *observation visits to other schools*, *teacher networks* and *mentoring or peer observation and coaching*. Notably, the three activities – *research activities*, *education conference/seminars* and *teacher networks* – distinguish English teachers from Chinese teachers significantly in that the latter have much higher participating frequencies.

During the past 18 months, most Chinese teachers have read professional literature, whereas over two-thirds of English teachers have not done so. Chinese teachers feel a higher level of benefit in the reading, whereas English teachers perceive a lower level of such benefit. During this period, all the 19 teachers but one have engaged in informal dialogue with their colleagues on how to improve their teaching, perceiving a moderate-to-large impact of this upon their professional development.

In six out of the 11 surveyed aspects of professional development, the teacher questionnaire sees little difference of development needs between teachers from the two countries. Nonetheless, Chinese teachers demand significantly higher needs than English teachers in five other aspects: (1) *classroom management*, (2) *knowledge and understanding of teachers' main subject field*, (3) *developing pedagogical content knowledge*, (4) *pupil discipline and behaviour problems* and (5) *pupil counselling*.

It is surprising that Chinese teachers wish for more development in both the subject matter and pedagogical content than English teachers when the literature suggests that Chinese maths teachers demonstrate profounder knowledge than their Western counterparts (for example than American teachers, see Ma, 1999). Moreover, despite having better disciplined classrooms than the English (as shown in Chapter 6), Chinese teachers demand higher needs than their English colleagues in developing management and disciplining skills, which is interesting to see.

Teacher school-based activities

Teacher responses about school-based activities in the teacher questionnaire shed light on the way in which different teachers are organised within each country and in which school-level policies indirectly lead to the differences of learning outcomes between countries. English teachers participated significantly more frequently than their Chinese colleagues in seven categories of activities:

- attending staff meetings to discuss the vision and mission of the school (#30a),
- developing a school curriculum or part of it (#30b),
- discussing or deciding on the selection of instructional media (#30c),
- exchanging teaching materials with colleagues (#30d),
- ensuring common standards for evaluations for assessing pupil progress (#30f),
- engaging in discussion about the learning development of specific pupils (#30g),
- discussing and coordinating homework practice across subjects (#30*l*).

This is reasonable if we take each country's context into consideration. China is less democratic and has greater power-distance than the case of England, hence less involvement of teachers into school management and decision-making, such as the activities #30a and #30f. Moreover, England doesn't unify textbooks and teaching materials at all levels, whereas China, although in a trend of decentralisation, only allows the diversity of textbooks at the provincial level, i.e. within each province textbooks are still unified. This explains the fact that English teachers tend to take part in activities #30b, #30c and #30d more frequently than Chinese teachers who do not have to do so because everything seems already decided beforehand. Inter-discipline is largely a rare fashion in schools in China; also, Chinese primary teachers are specialists who mainly focus on one specific subject; English primary teachers are generalists who teach almost every core subject in a class. These three facts make Chinese teachers less likely to coordinate cross-subject activities than their English counterparts.

According to the teacher questionnaire results, Chinese teachers get involved more frequently than English teachers into four categories of activities:

- teaching jointly as a team in the same class (#30h),
- taking part in professional learning activities (#30i),
- observing other teachers' classes and providing feedback (#30j),
- engaging in joint activities across different classes and age groups (#30k).

For a Chinese class, there are a number of teachers teaching different subjects and a class tutor (in Chinese 班主任, literally 'the class director') who often teaches Chinese or mathematics and plays a pastoral role. The class tutor is also responsible in coordinating collaborations between different subject teachers. For a primary class in England, there is often just a teacher and a TA who most of the time does what the teacher asks her/him to do to support low-ability or SEN children. English teachers seem more isolated than their Chinese counterparts who share close connections to the same class with a number of colleagues. It is perhaps because of the facts stated above that the two countries' teachers rate the frequency of activity #30h considerably differently.

Frequent peer observation and feedback among teachers is also a common phenomenon in Chinese schools; there are also frequent activities in which a group of colleagues (teaching the same subject within and/or across age groups) and internal and/or external leaders observe a lesson together and then jointly discuss the quality of the lesson and possible improving plans immediately afterwards. Within Chinese schools, there are also subject-based and age-group-based teaching research groups in which regular (often weekly) meetings are held for teachers to discuss strategies of improving teaching and learning quality and tackling specific issues which have recently arisen. It is thus not surprising to see the much higher participation frequencies of #30i and #30j rated by Chinese teachers than by English teachers. Based on similar reasons, the difference in #30k also seems understandable. In Chinese schools, teaching is more of a collective wisdom than an individual mission; in English schools, teaching is more of a teacher's individual exploration than a joint effort.

Teacher beliefs – direct transmission or constructivist?

Question #29 in the teacher questionnaire is intended to find out whether teachers hold "direct transmission beliefs" or "constructivist beliefs" about teaching and learning (OECD, 2009a, p. 92). Both are surveyed with four items (four-point scales ranging from strongly disagree to strongly agree). The former contains statements such as 'effective teachers demonstrate the correct way to solve a problem (#29a)', 'instruction should be built around problems with clear answers and around ideas that can be grasped quickly (#29g)', 'teaching facts is necessary as learning needs background knowledge (#29h)' and 'a quiet classroom is generally needed for effective learning (#29k). The latter belief is paraphrased as 'my role as a teacher is to facilitate pupils' own inquiry (#29d), 'pupils learn best by finding solutions to problems on their own (#29f)', 'pupils should be allowed to think of solutions to practise problems themselves before being taught (#29i)' and 'thinking and reasoning processes are more important than specific curriculum content (#29l)'.

The general stereotype is that Chinese teachers must fall in the direct-transmission category, and the English should absolutely be constructivist. There are indeed three direct-transmission aspects (#29a, #29h and #29k) agreed or strongly agreed by a higher percentage of teachers in China than in England, where, however, 50% or 80% of the teacher participants also agree to #29a or #29h; the aspect #29g seems agreeable to a similar proportion of Chinese and English teachers. Nonetheless, in the four constructivist aspects, the proportion of Chinese teachers that vote positively for #29f and #29*l* is double that of English teachers; similar percentages of teachers from both countries achieve an agreement on the aspect #29d; all teachers agree on the aspect #29*i*. Thus, overall, it seems that the English mind is not necessarily direct-transmission free and that the Chinese mind tends to be more pro-constructivist than the English.

Subject(s) that teachers teach

All English teachers teach three or more subjects including maths and other core subjects; six Chinese teachers only teach one subject, i.e. maths; the other three Chinese teachers, whilst teaching maths, also teach a non-academic subject (Arts or PE) which often occurs once a week in a Chinese school.

Time arrangement in the maths class

On average, English teachers allocated 3.6% of class time for administrative tasks, 8% of the time for keeping order in the class and 88.4% for actual teaching and learning. The Chinese proportions are 0.63%, 6.13% and 93.25%, respectively. In every time slot, England has a larger standard deviation than China. Both countries have exceeded the TALIS 2008 international average (80%) of class time on teaching and learning (OECD, 2009a). In addition to the arrangement of class time, teachers are also asked to rate the frequencies of given activities over a school year. English teachers carry out more frequently five activities in their lessons over a school year, including explicitly stating learning goals (#38b), differentiation by ability (#38e), checking pupils' exercise books (#38i), pupils evaluating and reflecting upon their own work (#38l) and pupils working in groups based upon abilities (#38n). It should be pointed out that English teachers have zero standard deviations across their ratings on the first four activities.

Teacher self-perceived influence and social status

All teachers from both countries believe that they make a significant difference in pupils' lives and that they are successful with pupils in their classes. They all think that, in their schools, pupils and teachers get on well with each other, that pupils' well-being is regarded as important and that teachers are interested in children's voices. There are also some international differences. English teachers are more likely to feel well respected in their communities and are more confident in dealing with pupils, particularly those difficult ones, than Chinese teachers, which is interesting to see, given the high status of teachers often found in CHC societies.

Chapter conclusion

The chapter presents pictures of the participants as captured by the pupil and teacher questionnaires. With the knowledge about the participants and their backgrounds in mind, we now are going to view the results and findings of the study in the following two chapters.

6 Evaluating and correlating maths teaching and learning

Chapter overview

- Quantity of maths teaching
- Quality of maths teaching
- Outcomes of maths learning
- Effects of maths teaching

As aforementioned, using two observation systems, we have systematically looked into the quantity (measured with OTL) and quality (measured with ISTOF) of teaching; with 40 items from TIMSS 2003 (Grade-4 Maths), we have assessed pupil performance at two points in time; through statistical analysis, we have correlated teaching with learning. This chapter presents detailed results.

Quantity of maths teaching

As explained in Chapter 4, for each teacher's lesson, two types of time-related percentages were calculated to show time on five categories of classroom activities and pupil time on task.

In terms of the duration of a lesson, Chinese lessons ranged from 36 to 45 minutes, whereas English lessons lasted longer, ranging from 47 to 66 minutes, which were often followed by a period for another subject (mostly English and taught by the same teacher) without even a short break in-between.

OTL 1 – whole class interaction

Whole class interaction happens when the teacher interacts with pupils in the whole class through two-way communication – usually in the form of questioning/answering or discussion. It is different from another OTL activity, *whole class lecture* which is one-way communication from the teacher to the class.

As shown in Table 6.1, the number of *whole-class interactive* activities varied significantly across countries. Chinese teachers not only spent more percentages of lesson time on *whole class interaction*, but also tended to split the time into more chunks within shorter duration of a lesson. Roughly, the numbers of

Table 6.1 Number of whole class interaction segments in lessons

EN1	*EN2*	*EN3*	*EN4*	*EN5*	*EN6*	*EN7*	*EN8*	*EN9*	*EN10*
3	1	6	3	9	3	5	3	3	3
CN1	*CN2*	*CN3*	*CN4*	*CN5*	*CN6*	*CN7*	*CN8*	*CN9*	
7	6	6	10	5	10	7	7	6	

whole-class interaction segments in Chinese lessons doubled those in English lessons, except for Teacher EN3 and Teacher EN5's lessons.

Proportionally, all Chinese teachers spent more time on *whole class interaction* than English teachers. The frequency analysis aided by SPSS provides more details. The corresponding percentage in 10 English classes ranges from 12% to 47%, with a mean of 23.8% and a standard deviation of 11.6%; the percentage in 9 Chinese classes ranges from 56% to 89%, with a mean of 72.2% and a standard deviation of 9.3%. Across all 19 participating classes, the average percentage of time on *whole class interaction* is 46.7% and the standard deviation 26.9%.

It is obvious that intra-country differences are smaller than the inter-country differences ($t(17) = 9.97, p < 0.001$), which once again indicates the importance of conducting cross-national TER studies – to see the complete picture. The effect of nationality on the proportion of time allocated for whole class interaction is strong ($d = 4.63 > 1$).

OTL 2 – whole class lecture

Whole class lecture happens when the teacher is delivering information directly to the class without pupils' active participation and responses. Table 6.2 shows the numbers of *whole class lecture* segments in all classrooms. It is apparent that only seven of England's lessons had such activities.

The striking finding is that Chinese teachers did not lecture to the whole class at all across the nine Chinese classes, whereas seven out of 10 English teachers chose to do so during a period of 3% to 9% of their lesson time. England's average percentage for *whole class lecture* is 3.9%, with the standard deviation being 3.2%.

OTL 3 – individual/group work

How often a teacher used *individual/group work* also varied both within and across countries. Table 6.3 shows the number of segments coded into this type of activity in each lesson.

Moreover, though all teachers in both countries have arranged time for pupils to work individually or in pairs mainly on written tasks, the percentages of lesson time for this type of activity varied significantly both within England and between two countries. The international variation can also be reflected by the means and standard deviations calculated at the national and international levels: England has a mean of 46.6% ($SD = 24\%$); China has an average

Table 6.2 Number of whole class lecture segments in lessons

EN1	*EN2*	*EN3*	*EN4*	*EN5*	*EN6*	*EN7*	*EN8*	*EN9*	*EN10*
1	2	1	2	4	1	0	1	0	0
CN1	*CN2*	*CN3*	*CN4*	*CN5*	*CN6*	*CN7*	*CN8*	*CN9*	
0	0	0	0	0	0	0	0	0	

Table 6.3 Number of individual/group work segments in lessons

EN1	*EN2*	*EN3*	*EN4*	*EN5*	*EN6*	*EN7*	*EN8*	*EN9*	*EN10*
7	3	6	5	11	4	8	3	2	7
CN1	*CN2*	*CN3*	*CN4*	*CN5*	*CN6*	*CN7*	*CN8*	*CN9*	
5	6	5	9	4	9	6	6	6	

of 27.8% (SD = 9.3%); pooling both countries' data together, the international mean is 37.7% (SD = 20.4%). The international difference is statistically significant ($t(12) = -2.3, p = 0.04 < 0.05$) and the nationality of each class has a strong effect on its proportion of lesson time on independent/group work ($d = 1.13 > 1$).

OTL 4 – *classroom management*

Classroom management does not directly involve subject matter teaching, and it often occurs when the teacher is organising pupils and/or other materials for either activity preparation or discipline purposes. In the EMT project, eight out of 10 English teachers had spent time on this type of activity; none of Chinese teachers had done so (Table 6.4).

This type of activities occupied on average 3.5% (SD = 2.4%) of English lesson time. The discipline of Chinese classes was better than that of English classes, and the transition between activities were also speedy and natural in Chinese classes. It thus seemed unnecessary for Chinese teachers to 'manage' activities.

OTL 5 – *partial class interaction*

Partial class interaction is the type of activities in which the teacher is conducting two-way interactions with individuals or part of the class rather than the whole class. This was a typical phenomenon only found in the English classes where setting by ability was commonly practised; this type of interaction did not happen at all in any Chinese classes (Table 6.5).

In the English classroom, this type of activities often happened on the carpet in front of the IWB. In most cases, the teacher's purpose of interacting with the partial class was to re-teach low-ability pupils the content that had been taught by the teacher but not well understood by some pupils; often during the

Table 6.4 Number of classroom management segments in lessons

EN1	*EN2*	*EN3*	*EN4*	*EN5*	*EN6*	*EN7*	*EN8*	*EN9*	*EN10*
3	0	0	1	1	2	2	2	3	3
CN1	*CN2*	*CN3*	*CN4*	*CN5*	*CN6*	*CN7*	*CN8*	*CN9*	
0	0	0	0	0	0	0	0	0	

Table 6.5 Number of partial class interaction segments in lessons

EN1	*EN2*	*EN3*	*EN4*	*EN5*	*EN6*	*EN7*	*EN8*	*EN9*	*EN10*
3	0	0	1	0	5	4	2	1	4
CN1	*CN2*	*CN3*	*CN4*	*CN5*	*CN6*	*CN7*	*CN8*	*CN9*	
0	0	0	0	0	0	0	0	0	

partial class interaction time, the rest of the class were working independently on worksheets.

Most of the time, English teachers did not ask pupils to come forward to the carpet; pupils actively chose to come whenever they felt frustrated. In very rare cases, the teacher's purpose was to teach more advanced content to the few high-ability pupils. Overall, in English classes, the average percentage of lesson time on partial class interaction is 22.3% (*SD* = 25.4%) of lesson time.

Now that we have a picture of how lesson time was distributed into five different types of classroom activities in each of the 19 classrooms, we will carry on to see the percentages of pupils on task during even intervals of the lesson time.

OTL 6 – pupil time on task

The result shows that English classes in general have lower time on task than do Chinese classes. Of all 10 English classes, just one has all pupils on task throughout the lesson, whereas seven out of nine Chinese classes have no pupils off task throughout the lessons. The average percentage of pupils on task within England was 92.8% (*SD* = 4.8%), and the national average for China was 99.7% (*SD* = 0.7%), with the international mean being 96.1% (*SD* = 4.9%). The relationship between nationality and pupil time on task was statistically significant with a strong effect ($t(9) = 4.5, p = 0.001, d = 2.5 > 1$).

Summary on the OTL measures

OTL measured the quantity of teaching factors within the frame of lesson time. Overall, as shown in Figure 6.1, English teachers allocated almost half of the lesson time for pupils working individually or with their peers on exercises, whereas in about three-quarters of their lesson time, Chinese teachers were interacting with the whole class. English teachers allocated a varying proportion of lesson time to each of the five types of activities, whereas in all Chinese

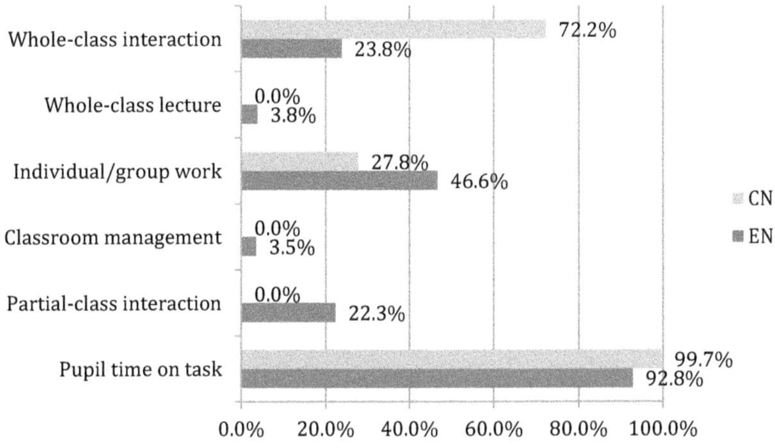

Figure 6.1 The international comparison of six OTL percentages

lessons, there were only two types: about three-quarters of whole-class interaction and about a quarter of individual/group work. On average, Chinese classes had a higher proportion of pupils on task, in comparison with English classes. In fact, only one pupil in each of two Chinese classrooms was off task over a short period, whereas only one English class had 100% pupils on task throughout the lesson. Chinese teachers did not appear to take actions to manage the class but were quite capable to keep children on task; English teachers used an average of 3.5% of lesson time to manage the class but had more children off task.

Chinese teachers on average put a bigger proportion of lesson time into whole-class interactive teaching and at the same time kept more pupils on task than their English counterparts. Existing TER evidence does suggest the positive connection between whole-class interactive teaching and pupil time on task (Croll, 1996; Muijs & Reynolds, 2003; Pollard et al., 1994; Teddlie & Liu, 2008). In a whole-class interactive lesson, the message is clear – it is *all* individuals, instead of one or some individuals, that matter at all times when the teacher includes everyone in the teaching and learning process. The highly individualised learning experience in the English classrooms echoed the ORACLE study (Galton & Simon, 1980; Galton, Simon, & Croll, 1980) which saw pupils working independently in two-thirds of the lesson time. It seems that over the past three decades or so, the proportion of English lesson time on independent work had decreased by about 20%, and that of time on whole-class interaction doubled. There was indeed change but just too small for about a third of a century. Progressive education – or rather the Western-perceived progressive education – has apparently maintained a lasting effect upon how teachers teach and children learn in the British classroom since the Plowden Report (Central Advisory Council for Education, 1967).

Both American and British bodies of literature suggest that whole-class interactive teaching works positively upon learning outcomes and gains (Emmer et al., 1979; Good & Grouws, 1979; Muijs & Reynolds, 2003; Pollard et al., 1994; Teddlie & Liu, 2008). We will see how teaching correlates with learning in the EMT study in later sections of this chapter.

Quality of maths teaching

Now let's look at the quality of teaching indicated by teacher scores in the six ISTOF domains (19 behavioural indicators). In each domain, we will first see a summary of the results of the measurement and then look in detail at the quality of teaching reflected by each behavioural indicator.

ISTOF 1 – assessment and evaluation

This component contains four scales, each ranging from 1 (strongly disagree) to 5 (strongly agree), which makes the full score of this domain 20. The average score of England is 12.8 ($SD = 2.3$), and that of China 19.4 ($SD = 1$), with the international mean score being 16 ($SD = 3.8$). The international difference is statistically significant ($t(17) = 8, p < 0.001$) with a strong effect ($d = 4 > 1$).

ISTOF 1.1 – did the teacher give explicit, detailed and constructive feedback?

Regarding answers or solutions to given problems, English teachers tended to state what was right rather than further scrutinise why it was right. Chinese teachers tended to facilitate pupils themselves to think and find the answers and/or a variety of solutions if possible and also to promote them to think why some pupils made it correctly whereas others did it incorrectly and how to avoid similar mistakes afterwards. They tended to prepare and carry out carefully the teaching arrangement between posing a question and achieving an agreed answer/solution, so that pupils' cognition was developed step by step towards realising the answer/solution themselves. Therefore, a clear distinction between English and Chinese teachers' approaches to questioning was who ultimately found the answer and told the truth – pupils or the teacher?

In responding to pupils' answers, English teachers' feedback was often nice and simple, and they seemed to hold the position in telling the truth and making formal judgement in the class; Chinese teachers tended to ask the rest of the class to judge whether a pupil's answer was correct or wrong; they were more likely to follow a pupil's answer with another question – often deeper than the prior question and at the same time building on the prior answer.

ISTOF 1.2 – was assessment aligned with goals and objectives?

All teachers from both countries had provided their pupils' assignments that were clearly related to what they learned. However, English teachers tended to stop, once learning objectives introduced and related assignments given; Chinese teachers focused more on the way in which assignments were connected with the learning goals.

ISTOF 2 – clarity of instruction

This component contained six items with a total score of 30. The English average is 16.4 (SD = 4.4), the Chinese average 27.9 (SD = 1.6), and the international mean 21.8 (SD = 6.7). Statistically, the international difference is significant, and the relationship between nationality and teacher scores in this domain is strong ($t(12) = 7.7, p < 0.001, d = 3.8 > 1$). Detailed explanations are as follows.

ISTOF 2.1 – did the teacher show good communication skills?

English teachers sometimes checked for understanding, but other times they simply carried on and dragged pupils' mind through the procedures they predefined for solving a certain type of problems, no matter whether pupils got it or not. In most classrooms, pupils did not get it at first, and the teacher had to re-teach the content multiple times. Chinese teachers asked lots of questions and sought their awareness of pupils' status of understanding through their answers. Overall, Chinese teachers communicated with the class in a clearer and more understandable manner than did their English colleagues.

ISTOF 2.2 – did the teacher give clear explanation of purpose?

All teachers in both countries clarified the lesson objectives during the first five to 10 minutes. There were, however, some international differences: English teachers all gave the learning objectives – so called *Success Criteria* – at the very beginning of the lesson; Chinese teachers all posed real life problems/questions and got pupils engaged in a whole-class discussion which eventually yet smoothly led the class to the learning topic. In addition, there were more reasoning activities going on in the Chinese classroom than in the English. Chinese pupils got more opportunities to think actively about the interconnections between the importance of the learning targets, the relevant real world issues, their prior knowledge and the given tasks; English pupils had more chances to think relatively narrowly that everything they did was mainly (if not solely) for the purpose of reaching the steps in their *Success Criteria*.

Moreover, the contents in the Chinese classrooms were almost all put in real-life contexts. For example, the first part of the lessons by teacher CN1 is about teaching and learning how to sort data and build a neat table of pupils' heights to order school uniforms for them. Even in lessons about knowing higher level numbers, teachers (e.g., CN4, CN6, CN7 and CN8) all arranged materials, such as the number of seats in Nanjing Olympic Stadium, for children to apply their newly learned knowledge and see mathematics in the world they live.

However, very few English teachers, for example, EN1 and EN2, did ask pupils to think about what they could go on to use the knowledge, for example times tables, and in either case, pupils did say that they could use times tables in future when shopping or doing other money- or number-related activities as adults, but throughout the teaching and learning process, the mathematical contents were hardly connected to or embedded in real life issues. The English lessons focused almost solely on pure mathematics.

ISTOF 2.3 – were lessons well structured?

English teachers applied relatively less time in formal instruction of new knowledge, and the rest of the time they were mainly re-teaching the same content to individuals or various small groups of individuals. During the formal whole-class teaching period (quite short), some (i.e. not all) English teachers tended to present the lesson content in a logical flow moving from simple to more complex concepts, despite the lack of deeper reasoning opportunities; during the majority of the lesson time, almost all teachers were busy re-teaching what they had just taught strictly following the same procedure as for the first input; the transition between lesson components was rough and lacked coherence.

Chinese teachers spent almost two-thirds of the lesson time gradually developing children's cognition from old to new knowledge; the strong logical flow was common in the Chinese classroom; the construction of mathematical knowledge was steady and coherent moving from simple to complicated concepts and/or varying situations; the transition between lesson components was natural and seemingly necessary at each specific time point because everything had been readily built for the transition; the smooth transition kept the momentum of teaching and learning from the beginning throughout the lesson to the end.

Overall, most English lessons were partially well structured during the period of whole-class first input and partially poorly structured during independent work periods when re-inputs to various individuals/groups were constantly occurring; Chinese lessons were very well structured throughout the lesson time.

ISTOF 3 – instructional skills

On the five 5-point scales of this ISTOF domain, the English score ranges from 7 to 14 (*mean* = 11.4, *SD* = 5.1), and the Chinese from 20 to 25 (*mean* = 23.2, *SD* = 1.7). The international mean is 17 (*SD* = 7.1). Again, the nationality poses a significant impact on teacher scores in this domain with a strong effect ($t(17) = 6.6$, $p < 0.001$, $d = 3.5 > 1$). The following paragraphs provide detailed accounts of teaching in two countries regarding the three indicators in the domain.

ISTOF 3.1 – was the teacher able to engage pupils?

The main methods of engaging children in learning in both countries were teacher questioning and classwork. There are, however, differences in this respect from country to country. In terms of questioning, Chinese teachers not only asked more questions but also asked different pupils to offer supplement/alternative answers to a single question, which helped engage pupils more deeply into mathematical thinking; English teachers asked fewer questions in a lesson, and they tended to keep a question to each individual, building little connection between individuals' answers. Furthermore, Chinese teachers were very unlikely to tell the correct answer to a given question, and they tended to let the correct answer emerge from pupils in the wake of intensive whole-class discussion. Nonetheless, in the English lesson, if the correct answer did not come out from the pupil being questioned, the teacher at the utmost asked two more pupils to have a go. After that, the teacher was ready to tell the correct answer/solution and explain detailed *standard* procedures, which neither posed enough challenge to children nor allowed them to think actively out of their comfort zone.

In the Chinese classroom, the assignments were tailored appropriately for the whole class in terms of difficulty levels and timing, and the lesson content was organised along the line of examples in a sequence such that, as time went on, the difficulty level and depth of knowledge gradually increased without children noticing the increase. Following each example, the teacher would ask pupils to work on a similar task independently. Each independent task was given at the time point when the class had collectively come up with one or multiple solutions to the example. Because of the thoroughness of the prior whole-class questioning and answering on the example (about 10 minutes), pupils were ready to complete the given task both quickly (often 1–2 minutes per task) and accurately. The assignments in the English classroom were also tailored appropriately but were strictly in line with the *Success Criteria* rather than with children's cognitive readiness; the assignment time came, in most cases, too early to allow teachers to make the new knowledge thoroughly absorbed by pupils beforehand. This also explained why English pupils kept coming to the carpet for repetitive teaching during independent work. They could not apply the new knowledge before they had fully understood and firmly grasped it.

Chinese teachers were capable of engaging the whole class into deep learning of the intended knowledge throughout their lessons, whereas English teachers failed to reach the depth of the academic engagement that they equally wholeheartedly wanted for their pupils.

ISTOF 3.2 – did the teacher possess good questioning skills?

English teachers tended to pose product questions which did not require complex reasoning and deep thinking. On the contrary, Chinese teachers asked more questions – mostly process questions which prompted deeper mathematical reasoning and thinking. The length of the pause following each question did vary a bit according to the difficulty level of questions in both countries' classrooms. Nevertheless, Chinese teachers tended to pause for an optimal length, and if the pause did not work, they would immediately pass the question on to other pupils or break the question into a series of sub-questions that were easier to answer, hence keeping the momentum of the lesson. Nonetheless, English teachers often gave the full explanation of the answer if a correct response did not emerge after one or two pupils had attempted the question; they seemed to believe it was part of their duty to immediately save the children out of these difficulties. On the contrary, when encountering similar situations, as explained in the indicator ISTOF 3.1, instead of providing/feeding the answers, Chinese teachers would keep inviting the next pupil to answer the question until one or a range of reasonable answers had arisen.

ISTOF 3.3 – did the teacher use various teaching methods and strategies?

English teachers tended to use simple teaching methods and follow fixed strategies which seemed to leave fewer surprises to the class, even though they all had amazing IWBs which could only be found in the classrooms of some commercial after-school courses in China. Chinese teachers, with less advanced equipment, however, were able to adjust their teaching methods and arrangements every now and then so that they could keep pupils *awake* and academically thoroughly engaged throughout a lesson; it appeared that the change of methods in the Chinese classroom was not in itself for the sake of change, but for the purpose of better developing children's certain aspect of knowledge at that specific point (in terms of time and cognitive development) in the lesson.

ISTOF 4 – promoting active learning and developing metacognitive skills

On the 10 5-point scales in this domain (4 indicators), English teachers attained an average of 21.7 (SD = 6.9), and Chinese teachers 47.3 (SD = 2.9). The international mean was 33.8 (SD = 14.1). There was a statistically significant and strong relationship between nationality and teacher scores in the domain ($t(17) = 10.3, p < 0.001, d = 5.2 > 1$).

In general, English teachers tended to ask product questions which led to exact facts or procedures, but Chinese teachers more often asked lots of process questions before and after a task to generate ideas as to how the task might be tackled or had been tackled and why. English pupils had to follow teachers' instruction of the *right* facts or procedures, whereas Chinese pupils had rich opportunities to plan and discuss strategies or reflect upon their and their peers' problem-solving experiences – either successes or failures. In the English class, there was often one set solution/procedure to a problem; in the Chinese class, solutions were often more than one and procedures could be flexible as long as they were logical and coherent.

ISTOF 4.1 – did the teacher help pupils develop problem-solving and metacognitive strategies?

English teachers asked pupils to use the strategies that they had taught them; Chinese teachers encouraged pupils to think of multiple strategies to solve a varying problem and share their strategies during whole-class discussion. English teachers were more likely to talk the pupils through the problem-solving steps than asking them independently to explain detailed strategies; Chinese teachers tended to ask pupils to explain various solutions they could come up with and to foster brainstorming discussions among them. Most teachers from both countries tended to provide instruction in problem-solving processes. Nevertheless, English teachers were likely to deliver the *right* strategies straightaway or after the first attempt of questioning, whereas Chinese teachers tended to ask a series of questions of varying depths to provoke children's thoughts so they could gradually find the strategies by themselves.

ISTOF 4.2 – did the teacher give pupils opportunities to be active learners?

All teachers from both countries encouraged pupils to ask one another questions and to explain their understanding of the topics to one another. English teachers tended to promote paired discussions; Chinese teachers inclined to whole-class discussions. All teachers would encourage pupils to correct their own work if they spotted pupils' mistakes. However, because there was lots of individual work parallel with seemingly constant re-inputs from the teacher to individuals or small groups on the carpet in the English classroom, English teachers did not have extra energy to capture all errors emerging in children's classwork, let alone asking them to correct each error. In the Chinese classes, pupils could see clearly what counted as correct and wrong solutions and why during whole-class interaction; during the independent work period, Chinese teachers all circulated fairly quickly through the class to check *every* pupil's work; they often picked up representative work samples – both correct and wrong ones – to share with the class via the projector immediately after each slot, asking pupils to make the judgement (right or wrong?), comments (how and why?) and corrections if needed.

ISTOF 4.3 – did the teacher foster critical thinking in pupils?

English teachers focused more on transmitting standard solutions (often just one fixed solution) rather than inspiring diverse solutions; they therefore did not frequently ask pupils to judge the pros and cons of different approaches. It was typical in the Chinese classrooms that multiple solutions were always applauded and pupil-led comparisons of different approaches were particularly encouraged. English teachers stopped scrutinising pupils' internal thinking processes once they made sure the correct solutions/procedures were well delivered from their own points of view. Chinese teachers, however, demanded deeper thinking amongst pupils; every task they gave would include pre- and post-task discussions; the latter nurtured post-action reflection on problem-solving, which seemed similar to action research. Despite differences, all teachers from both countries invited pupils to give their personal opinions on certain issues.

ISTOF 4.4 – did the teacher connect materials to pupils' real world experiences?

English teachers tended to discuss mathematical content with pupils for the sake of mathematics itself; Chinese teachers always started and ended their lesson in relevant phenomena in daily life and related the lesson content to the real world whenever necessary throughout the lesson. English pupils were not often invited to raise examples from their life experiences, and in particular there was a severe lack of opportunities for the children to make links between specific knowledge points and their real world experiences. Conversely, their Chinese peers were more frequently asked to share personal experiences that were closely related to the intended mathematical content throughout the lesson.

ISTOF 5 – classroom climate

The domain of classroom climate contains eight 5-point scales clustered around four indicators. The international average was 30.3 ($SD = 9.2$), with the English mean being 22.8 ($SD = 6$) and the Chinese mean 38.8 ($SD = 1.1$). The international difference was statistically significant, and the relationship between nationality and teacher scores was strong ($t(10) = 8.3, p < 0.001, d = 4.5 > 1$).

ISTOF 5.1 – were all pupils valued?

All teachers from both countries demonstrated genuine warmth and empathy towards and showed respect for the pupils in their classrooms in all respects.

ISTOF 5.2 – did the teacher initiate active interaction and participation?

Chinese teachers were more able to engage every pupil in productive work which generally took pupils much less time and was better completed than the English case. Most English teachers failed to do so and had to re-teach individuals and groups of pupils for multiple times what had been taught because pupils consistently encountered difficulties during independent work. Moreover, across classrooms, Chinese teachers asked more questions than their English colleagues, and Chinese pupils were given more opportunities to answer questions than their peers in England. Chinese teachers' questions tended to follow a coherent logical flow – one led to another, which nurtured deeper thinking amongst pupils; English teachers' questions tended to be dotted with little or loose connections and pose little challenge to pupils who were expected to passively learn the set solutions that the teacher had demonstrated.

ISTOF 5.3 – did the teacher interact with all pupils?

All teachers showed their effort in involving all pupils, particularly those who did not voluntarily participate in classroom activities. Nevertheless, English teachers applied more time interacting with individuals, which hence limited their degree of interacting with *all* pupils in a lesson. The Chinese approach, which featured a greater proportion of time on whole-class interactions, on the other hand, maximised frequencies of teacher-pupil and pupil-pupil interactions.

ISTOF 5.4 – did the teacher communicate high expectations?

Chinese teachers were more likely to pose a reasonable challenge – higher than pupils' actual level but achievable after a little effort – to pupils and briskly praise them for realising their potential. Though English teachers were generous in praising pupils, they often asked easy and less challenging questions and *fed* pupils with too many hints and tips while waiting for answers, thus lowering the degree of challenge they could have posed to them. The expectation from Chinese teachers to their pupils seemed much higher than the expectations from English teachers. Chinese pupils were better motivated, therefore being more confident to meet bigger challenges than their peers in England.

To summarise, the focus of the *classroom climate* domain is more about whether the teacher seeks to build the atmosphere where active learning is nurtured and pupils are motivated to attain more advanced knowledge and to seek deeper mathematical thinking. In this respect, Chinese teachers did better than English teachers.

ISTOF 6 – classroom management

In the domain of classroom management, there are seven 5-point scales under the umbrellas of three indicators. Teachers in England scored much less than their colleagues in China. The former got a mean of 18.1 (SD = 6.4), and the latter 34.2 (SD = 1.7). The difference of scoring between countries also partially explained why the Chinese classrooms were better disciplined. There was a statistically significant and strong relationship between nationality and teacher scores in this dimension ($t(17) = 7.3, p < 0.001, d = 3.98 > 1$). Detailed accounts are given in the following paragraphs with regard to each indicator.

ISTOF 6.1 – was learning time maximised?

Teachers all started their lessons on time. English teachers, however, were unable to ensure that all pupils were fully involved in learning activities until the end of the lesson, mainly because they had to stay at the carpet to provide possible re-inputs during the independent work time. Chinese teachers were conscious of the state of both individuals and the whole class and were able to keep teaching and learning processes more engaging and pupils on task. English teachers did not take many actions to minimise disruption – they seemed quite tolerant with a certain degree of chaos particularly during the periods of independent work. Neither did any Chinese teachers take such actions overtly, because they did not have to. They had arranged the teaching and learning process in a way that disruption had been covertly prevented.

ISTOF 6.2 – were clear rules evident?

This dimension looks at two aspects (i.e. 2 scales) of clear rules, clarity of getting help from the teacher and clarity of pupils' options after finishing assignments. First, in the English classrooms, the most efficient way of getting help was to go to the carpet and find the teacher; sometimes pupils had to queue up; a few of them sometimes raised their hands to call for help, but it seemed to take an even longer time for the teacher to spot their hands in the air. In the Chinese classrooms, teachers seemed widely alerted both visually and mentally and were ready to capture any moves in the class immediately; raising hands was therefore the most popular and quickest way for pupils to get attention and help from the teacher.

Second, in the English classes, there was a lack of clarity about what options were available after the pupils had finished their assignments; there was, however, clarity about what options were available when the pupils encountered difficulties – to approach the teacher on the carpet almost anytime they felt the need. In the Chinese classes, teachers always announced clearly what options were available to those who completed tasks earlier than others; sometimes the option was to check their answers with their desk-mates; sometimes it was to sit upright and check their own answers so the teacher could know straightaway that they were done by simply looking at their sitting gestures; a brief praise would be given to them if they completed earlier and still managed to behave well while checking their work.

ISTOF 6.3 – were misbehaviours and disruptions effectively dealt with?

After the short whole-class teaching time (i.e. the first input), English teachers were always busy interacting with individuals or a group of individuals on the carpet, so non-academic behaviours among pupils were often not spotted and dealt with. However, there rarely were any misbehaviours and disruptions in the Chinese classrooms; on very rare occasions (only twice across all lessons) when they occurred, Chinese teachers dealt with them instantly and smoothly. All classes from both countries seemed to have their own conventions and rules regarding intolerant misbehaviours and disruptions.

Having looked into details of the ISTOF results, we will now step back to see the whole picture that all ISTOF measures drew about the quality of mathematics teaching in two countries.

Summary on the ISTOF measures

Overall, for the whole instrument, English teachers scored less than Chinese teachers. English teachers achieved an average of 103.2 (SD = 29.2), whilst Chinese teachers had a mean of 190.9 (SD = 7.9). The relationship between nationality and the sum of teacher ISTOF scores was statistically significant and strong ($t(17) = 8.7, p < 0.001, d = 4.7 > 1$). As shown in Figure 6.2, at the international level, English teachers on average achieved lower scores than did their Chinese colleagues on ISTOF scales.

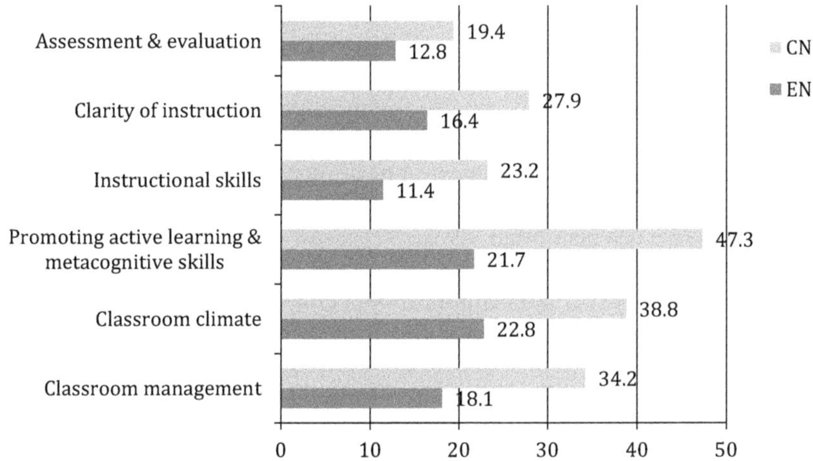

Figure 6.2 The international comparison of six ISTOF scores

Since the very early stage ofTER, instructional clarity has always been found to be a teaching factor that works effectively on learning (Clarke et al., 1979; Good & Grouws, 1979). The clarity should not only structure the observable part of teaching, such as classroom activities, but also run deep into the cognitive dimension of the lesson where the teacher explains the logical structure of knowledge, solutions, reasons behind wrong solutions or misconceptions and so on in such a clear manner that misunderstandings have little chance to settle in pupils' minds after a lesson. Such internal clarity of teaching would thus mean the clear instruction of new knowledge and a clear and coherent connection between the new and old knowledge whenever necessary.

In the English classroom, children kept coming back to the teacher for re-inputs, which should be read as a strong signal asking the teacher for a clear and accurate diagnosis of possible knowledge holes formed in the past. On those occasions, the English teacher should be able to immediately generate and implement a clear plan to bridge the knowledge gap with great clarity. For the case of mathematics teaching, teachers well aware of the hierarchical structure of the subject have a strong potential to build clear connections between different areas of mathematics, which leads to a clear instruction by the teacher and subsequently clear understandings amongst pupils in and across lessons (Askew et al., 1997).

Interacting with the whole class would demand a lot more instructional skills than interacting with individuals or a small group of pupils, since the former has to address a class of needs and engage all pupils simultaneously. Effective teachers have the kind of skills to break the knowledge down to small chunks and pose multiple interconnected questions which are more likely to generate correct answers from the class (Berliner et al., 1978; Brophy, 1973; Fisher et al.,

1978; Good & Grouws, 1977) through involving everyone in the whole-class brainstorming/discussion (Emmer et al., 1979; Evertson et al., 1980).

Teachers' *appropriate* responses to different types of answers or reactions to pupil behaviours also have positive effects on learning outcomes and gains. The key rule is to get the pupils to receive the message as to whether an answer is right or wrong or whether or not a conduct is acceptable without leaking too much teacher emotion, e.g. overpraise or heavy criticism (Anderson et al., 1979; Brophy, 1973; Soar & Soar, 1973).

More effective teachers often ask more process questions which offer more opportunities for the pupils to explain the thinking of thinking – metacognition (Brophy & Good, 1986; Munro, 1999). Frequent teacher questioning in the whole class means that every pupil is expected to be an active participant of the lesson. This way of teacher questioning in the whole class certainly sends a clear message to all pupils that everybody is able to come up with an answer, which thus gives higher expectations to all and forms an academically-encouraging classroom climate. Previous research does suggest that pupils who receive high expectations from their teachers are more likely to access wider curriculum coverage whereas low expectations are related to dropped attainment (Mortimore et al., 1988; Tizard et al., 1988).

Both OTL and ISTOF measure classroom management, but the former focuses on the quantity of it and the latter the quality. Our review of literature suggests an effective maths lesson has less time (quantity) spent managing the class and higher quality of management results, such as more time on academic teaching and learning and less pupils off task (Brophy & Good, 1986; Evertson et al., 1980; Good & Grouws, 1977). It therefore makes sense to measure the same type of teacher behaviour in these two dimensions.

With the teaching differences in mind, we will now look at the picture of children's maths performance measured at two points in time. The primary attempt is to assess both learning outcomes and learning gains over 10 school weeks, but the TIMSS 2003 Grade 4 maths items turn out to be too easy for the Chinese pupils to measure accurate learning gains. The ceiling effect thus means that only learning outcomes will be systematically analysed and further correlated with teaching factors. The subsequent section focuses on the effects of maths teaching on learning. Pooling data from two countries together, different teaching measures are correlated with learning outcomes at the classroom level. Effective and ineffective maths teaching factors are identified.

Outcomes of maths learning

In this section, test results are presented in three ways to generate insight from different angles:

- test results by item,
- test results by domain,
- overall test results.

Test results by item

Within each country, the correct rate of each item was calculated through dividing the number of children who did the item correctly by the total number of all that took the test. Test 1 saw the correct-rate gaps of 29 items between England (lower rates) and China (higher) ranging from 18 (inclusive) to 72% and such international gaps on 7 items ranging from 1 to 14% (Figure 6.3). While both countries had the same rates – 86% – on items #17, England outperformed China on items #9, #10 and #13 by 49%, 19% and 17%, respectively. It should be pointed out that the items #9 and #10 represented numbers with the Western-style tally marks. This might have caused the relatively lower correct rates of Chinese pupils who were used to the Chinese equivalent style – writing the five strokes of the Chinese character 正.

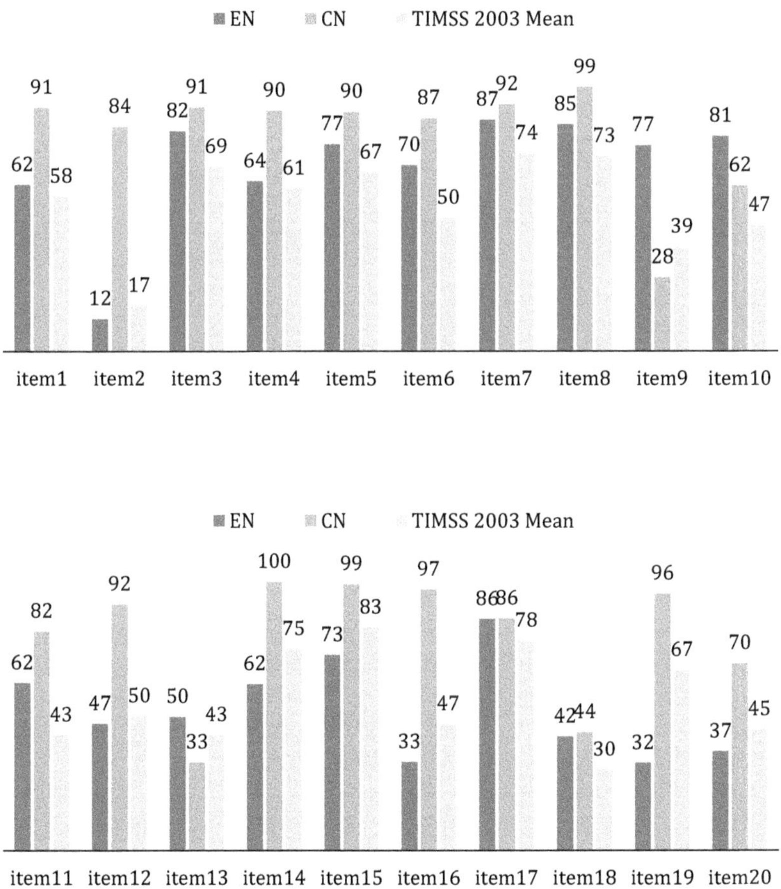

Figure 6.3 A comparison of item correct rates in Test 1 (%)

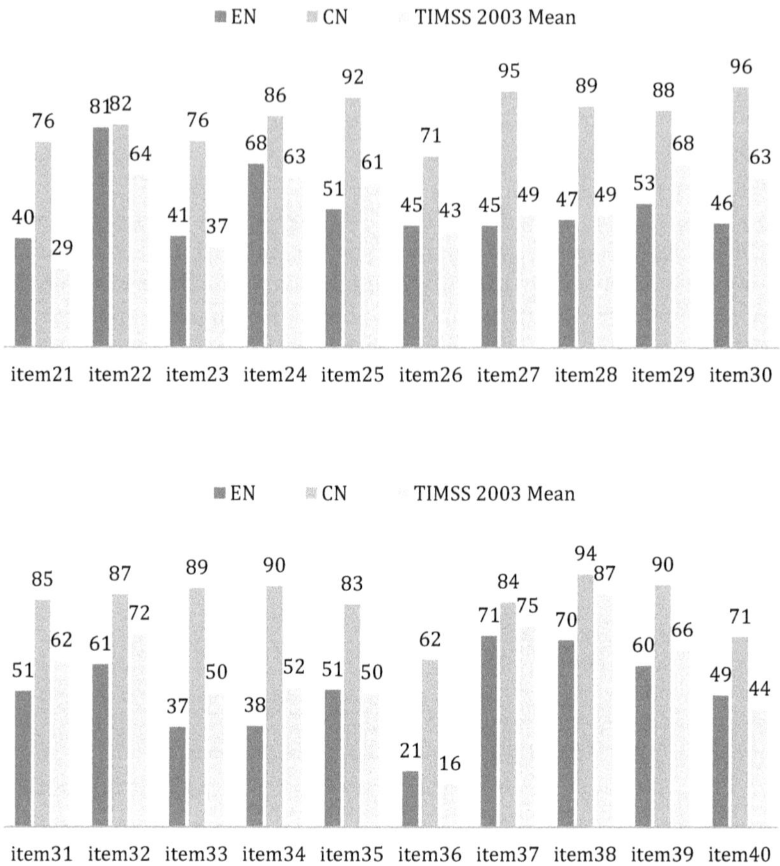

Figure 6.3 (Continued)

In comparison with the TIMSS 2003 average, English correct rates were higher on 21 items, with the difference ranging from 1 to 38%, but English rates were also 2 to 35% lower in the remaining 19 items. Under the same comparison, Chinese correct rates exceeded the TIMSS 2003 average by 18 to 67% on 31 items and 7 to 16% on 7 items and lagged behind the TIMSS 2003 mean by 10 and 11% on items #9 and #10, respectively, which, as aforementioned, involved the representation of numbers with Western-style tally marks.

In Test 2, English correct rates were 43, 18, 10, 1 and 1% higher than the Chinese ones on five items – #9, #10, #13, #17 and #22, respectively; Chinese correct rates were higher than the English ones on all remaining 35 items; the Chinese-English gap ranged from 18 to 62% on 22 items, and from 4 to 17% on 13 items (Figure 6.4).

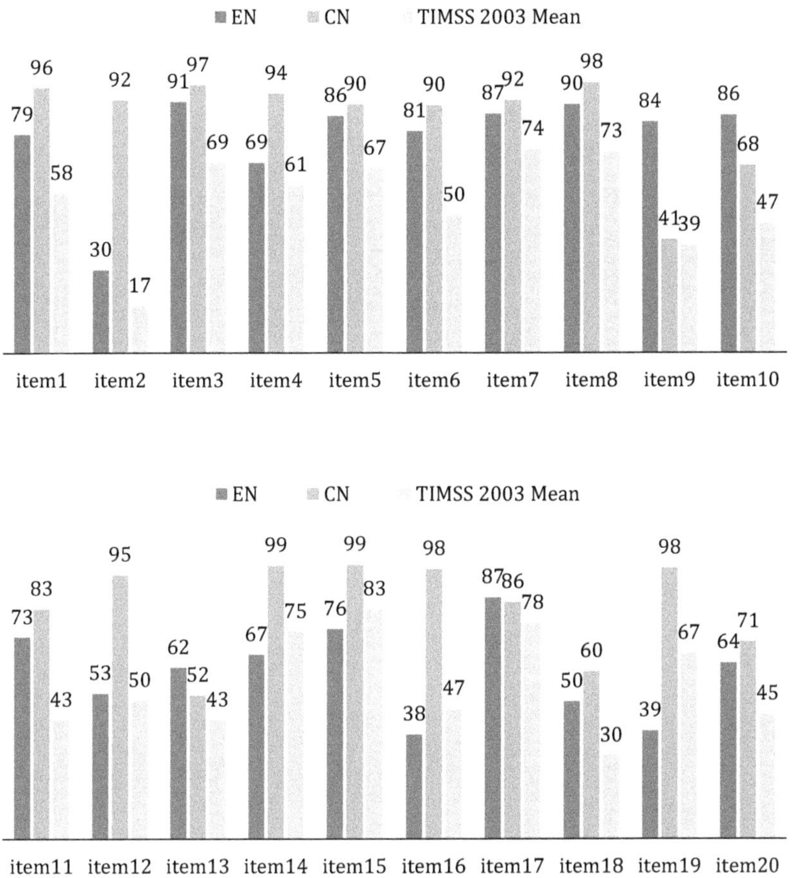

Figure 6.4 A comparison of item correct rates in Test 2 (%)

Vertically (or historically) comparing with the TIMSS 2003 average, English pupils had achieved 20 to 45% higher correct rates on nine items, 12 to 19% higher rates on 11 items, 1 to 9% higher rates on 11 items; they got, however, 4 to 28% lower rates on seven items and the same rates as the TIMSS average on two items – #31 and #37. Following the same comparison route, Chinese pupils achieved higher rates on all items of which they made 18 to 75% higher correct rates on 35 items, 8 to 16% higher on four items and 2% higher on item #9 than the TIMSS 2003 average.

Overall, Chinese pupils made much higher correct rates in comparison with either English pupils in this study or pupils on the average line in TIMSS 2003. The positive differences between the correct rates of England and TIMSS 2003

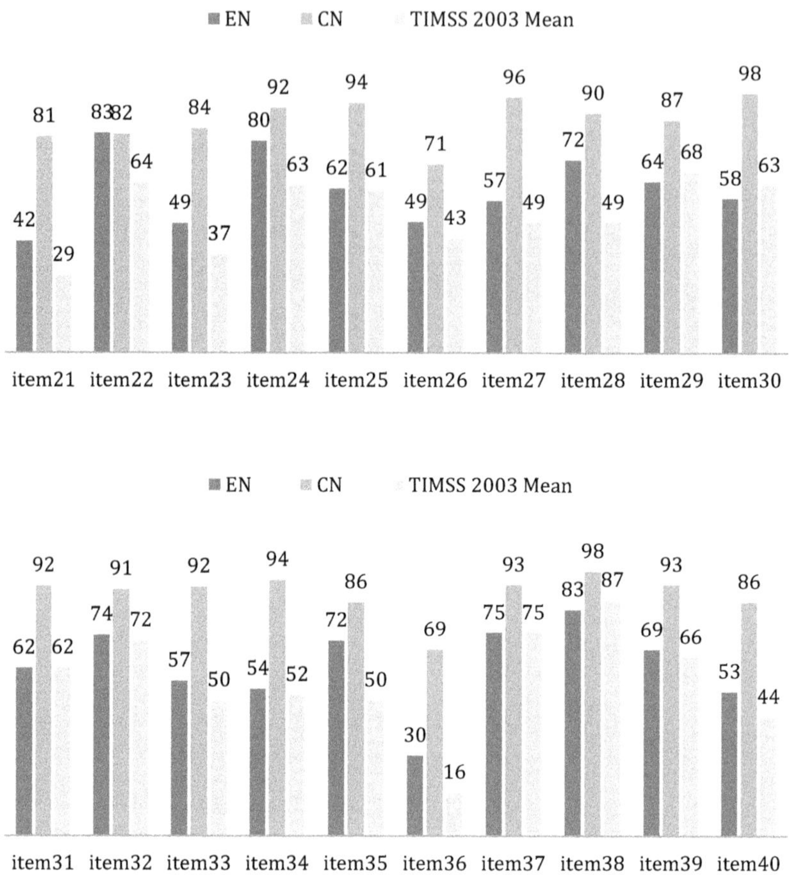

Figure 6.4 (Continued)

average, where there were any, tended to spread evenly across various ranges, whereas such gaps between Chinese correct rates and the TIMSS 2003 average skewed dramatically to the higher end.

Test results by domain

Having seen the results of the two tests by item, next we will look at the results by content domain and cognitive domain.

Defined by the TIMSS 2003 Assessment Framework (Mullis et al., 2003), each TIMSS item belongs to a specific content domain and a particular cognitive domain. We shall now leave the item level and zoom out a bit to compare the correct rate of each content domain and cognitive domain cross-nationally in each test.

In Test 1, in terms of correct rates, English pupils' best content domain was *Data* (Figure 6.5), and their best cognitive domains were *Solving routine problems* and *Using concepts* (Figure 6.6). Chinese pupils' strongest content domain was *Patterns and relationship*, and their best cognitive domain *Knowing facts and procedures*. Looking cross-nationally, we found, in all but one domain – *Data*, Chinese pupils achieved much higher correct rates than their English peers.

In Test 1, English pupils achieved the least correct rate in *Number* and *Knowing facts and procedures*; Chinese pupils had the lowest rate in *Data* and *Reasoning*.

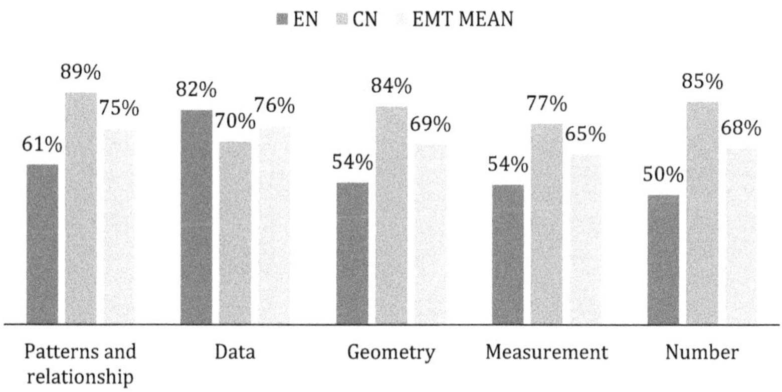

Figure 6.5 Correct rates by content domain in Test 1

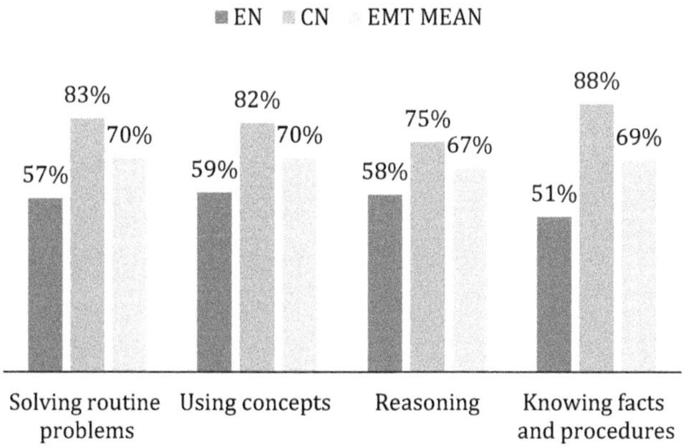

Figure 6.6 Correct rates by cognitive domain in Test 1

English pupils' correct rates were all below the EMT international average, except for the content domain of *Data* in which they exceeded the average by 6% and the Chinese correct rate by 12%. Apart from *Data*, English pupils were below the EMT average in any other content or cognitive domains by 9 to 19% and lower than the Chinese correct rates by 23 to 36% in four other content domains and by 17 to 37% in all four cognitive domains; Chinese pupils were above the EMT average by 12 to 19% in four other content domains and all four cognitive domains.

In Test 2 (Figure 6.7 and Figure 6.8), Chinese pupils again got the lowest correct rate in the domain of *Data* but much higher rates in all other domains.

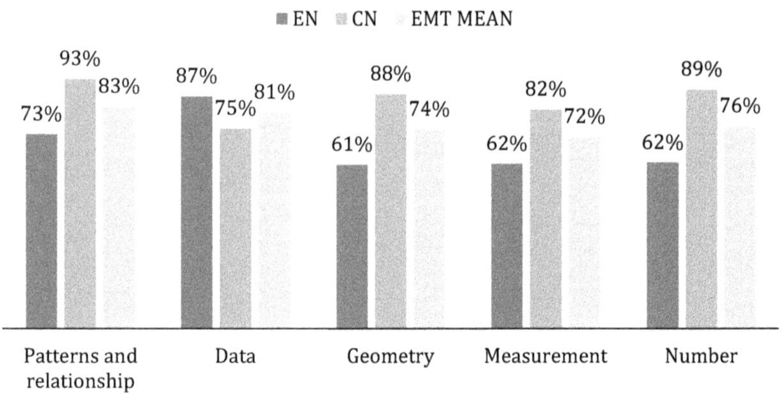

Figure 6.7 Correct rates by content domain in Test 2

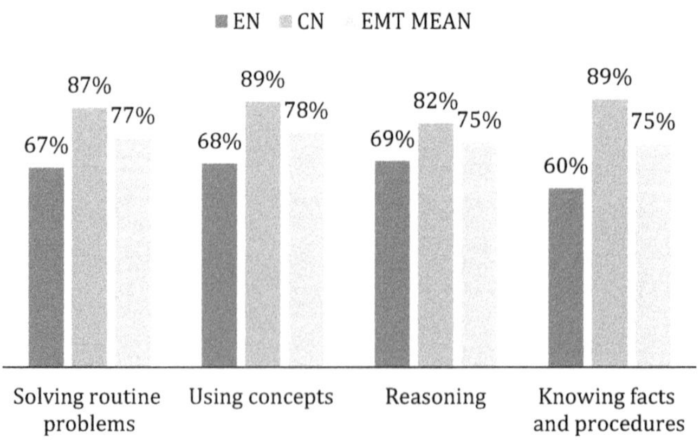

Figure 6.8 Correct rates by cognitive domain in Test 2

English pupils' strongest domains were *Data* and *Reasoning*, and Chinese pupils' *Patterns and relationship*, *Using concepts* and *Knowing facts and procedures*. Again, except for *Data*, English pupils attained 6 to 15% lower correct rates than the EMT average and 13 to 30% lower than their Chinese peers in all other domains; Chinese pupils got 6 to 15% higher correct rates than the EMT average in other eight domains.

Having scrutinised children's performance at the item and domain levels, we will now zoom further out to see the full picture – the overall test results.

Overall test results

Both of the two maths tests showed that Chinese pupils on average outperformed their English peers by over 20%. In Test 1, with a difference of 27, English and Chinese pupils had achieved a mean of 56 and 83, respectively. The within-country variance was quite different (England: $N = 231$, $SD = 21$; China: $N = 250$, $SD = 10$). The international difference was significant and the relationship between nationality and pupil performance was strong ($t(321) = -17.7, p < 0.001, d = 1.74 > 1$). In Test 2, on average, the English pupils have achieved 66 ($N = 236$, $SD = 19$), and the Chinese 87 ($N = 326$, $SD = 10$), which indicated a gap of 21 between the two countries on average. The former have gained approximately 10%, probably due to a quite low starting point and therefore sufficient space for improvement; the latter only made it to a 4% increase. The international difference was statistically significant and strong ($t(325) = -15.4, p < .001, d = 1.44 > 1$).

English pupils had made a progress of 9.7 over the 10 school weeks between two tests, whereas their peers in China only improved 4.2 on average. This does not mean that Chinese teachers had added less value to pupils' learning in mathematics over time than had their English colleagues. The Chinese means in the two tests were both about 10 points higher than the corresponding international means, whereas English means were over 10 points lower than the average on both occasions. It was assumed that there might be a ceiling effect on Chinese pupils in the second test that did not allow their scores to grow.

To find the proportions of pupils that had achieved a score that was close to or reached the ceiling, a line through the 60 percentile score was drawn to see how many pupils from each country had made it to the top 40% in the two tests. In Test 1, the test score separating the top 40% achievers from the rest was 80 (i.e. the 60 percentile score); in Test 2, the score was 86.3. Then, further calculations were made regarding the proportions of English and Chinese pupils that were among the top 40%, i.e. between the 60 percentile (80 for Test 1 and 86.3 for Test 2) and the ceiling (100).

Results are shown in Figure 6.9 and Figure 6.10. In either test, none of the English pupils had made it to the ceiling, but a small proportion of Chinese pupils had already achieved a full mark. In Test 1, 13.4% of English pupils and 42.7% of Chinese pupils were amongst the top 40% in this cohort of 9- to 10-year-olds. In Test 2, the proportions were 12.9% (EN) and 40.3% (CN),

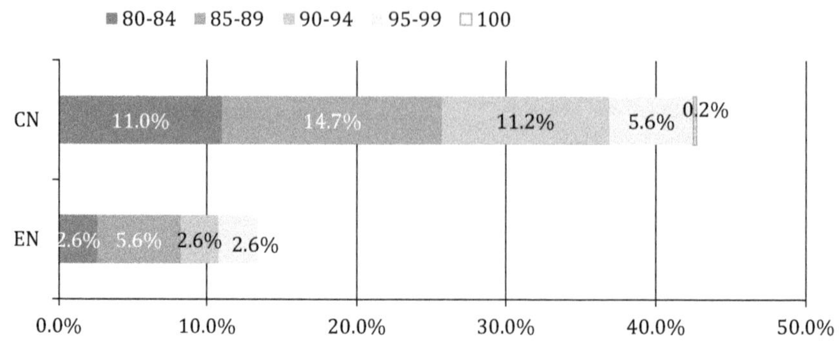

Figure 6.9 The proportions of top 40% pupils in Test 1

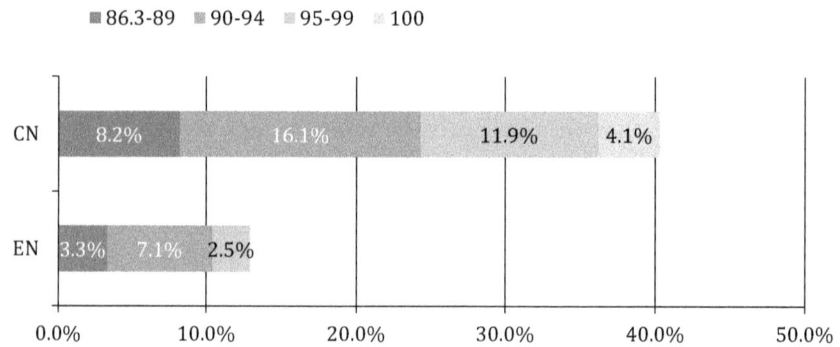

Figure 6.10 The proportions of top 40% pupils in Test 2

respectively. There is indeed a ceiling effect upon Chinese pupils which did not provide enough space for them to improve their scores between the two tests. This ceiling effect might be caused by the limit of the TIMSS 2003 items which were too easy for Chinese pupils. These items were apparently not suitable for testing learning gains in this study. It was thus decided that teacher behaviours should be correlated with pupils' maths performance at points in time, rather than learning gains over time.

The differences between the average scores of English pupils and Chinese pupils in the two tests are consistent with their performance differences measured in international surveys, such as PISA and IAEP (Lapointe et al., 1992; Lapointe et al., 1989; OECD, 2010, 2013).

Effects of maths teaching

Teacher scores in OTL and ISTOF were correlated with pupil performance in each test at the classroom level across two countries (Figure 6.11 and Figure 6.12).

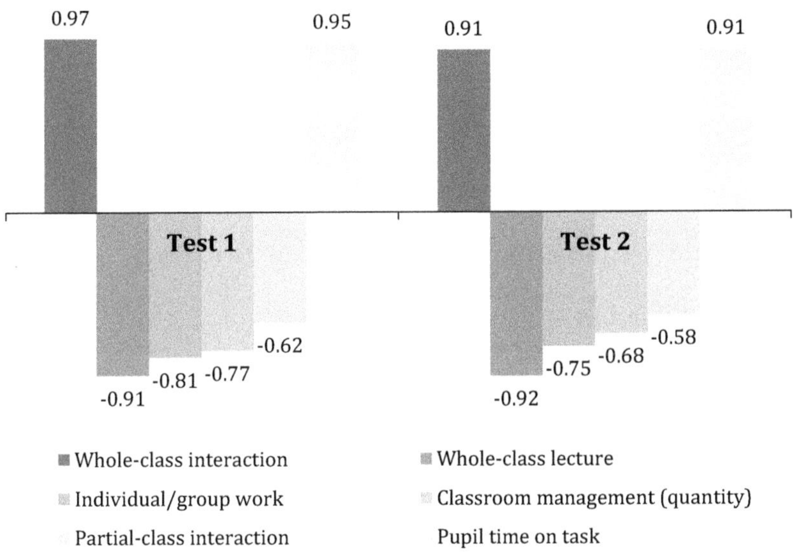

Figure 6.11 Correlations between OTL and learning outcomes

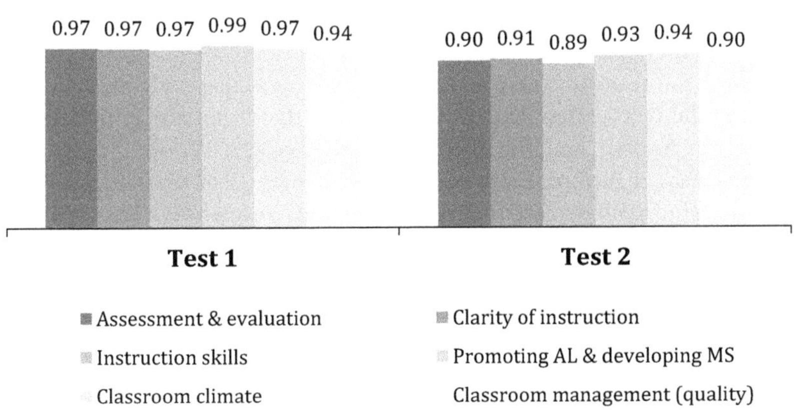

Figure 6.12 Correlations between ISTOF and learning outcomes

The pooled analysis suggested that eight teacher behaviours/factors were significantly (*p* at the 0.01 or 0.05 level) positively correlated with maths performance and three others had a negative impact. The effective teaching behaviours/factors include:

- whole-class interaction,
- pupil time on task,
- assessment and evaluation,
- clarity of instruction,
- instructional skills,
- promoting active learning and developing metacognitive skills,
- classroom climate,
- classroom management (quality).

Those negative and ineffective teaching behaviours/factors are:

- whole-class lecture,
- individual/group work,
- classroom management (quantity/frequency).

Partial class interaction might also have posed a negative effect upon pupil learning, but the impact was not statistically significant.

As reviewed earlier on, it is evident in the literature that whole-class interaction, time on task, teaching clarity, teacher questioning skills, high expectations and the quality of classroom management, as opposed to the quantity/frequency of classroom management, have a positive effect on attainment and/or achievement (Clarke et al., 1979; Emmer et al., 1979; Good & Grouws, 1979; Muijs & Reynolds, 2003; Pollard et al., 1994). In the TIMSS 1999 Video Study, more time on whole-class teaching was a typical feature of the Japanese maths lessons (Hiebert et al., 2003).

However, it is clear that whole-class teaching does not guarantee a positive effect, since the teacher could be either *actively interacting* with the whole class or *lecturing* the whole class. It is two-way, rather than one-way, communication between the teacher and the whole class that works. Whole-class lecturing, essentially teacher monologue, rules out the dynamics of teacher questioning, pupil answering, teacher reacting to pupil answers and whole-class discussion – the combined elements that promote thinking and deepen understanding and learning (Good et al., 1978).

It has to be pointed out again that, in the EMT study, the highly individualised teaching and learning in the English classroom does correlate with poor performance in maths. Despite the fact that enormous efforts have been put into altering the heavily individualised teaching towards whole-class interactive teaching in Britain since the end of last century (DfEE, 1998), the present data and results do not suggest that much change has occurred.

Chapter conclusion

This chapter shows large inter-country differences in the quantity and quality of teaching as measured with OTL and ISTOF and, in either measurement, a bigger intra-country difference in England than in China. This pattern applies to children's maths performance as well. In either maths test, English Year-5 pupils have a significantly lower mean score and a much wider gap within their nation than pupils from China of the same age. Pooling the teaching and learning data from all 19 classrooms together, we have evaluated the impact of maths teaching upon learning cross-nationally. The correlational results have suggested a list of effective or ineffective maths teaching behaviours/factors that may work positively or negatively upon children's learning in classrooms of both countries. These findings may be further extended to countries that have similar educational systems where similar effects of those teaching 'elements' may be expected.

7 Seeing maths teaching through different eyes

Chapter overview

- Seeing maths teaching through the researchers' eyes
- Seeing maths teaching through individual teachers' eyes
- Seeing maths teaching through groups of teachers' eyes

This chapter presents a jigsaw of mental pictures of the researchers, teachers, their colleagues and their foreign colleagues regarding the effectiveness of maths teaching. First, the researchers' unstructured observations tell the typical structure (*lesson flow*) and key features (*characteristics*) of mathematics lessons in each country. Second, interviewing data are presented and discussed to show teachers' perceptions of effective mathematics lessons, their own judgement of the quality of their observed lessons, their awareness of teaching in the other country and their attitudes and openness towards change in practice. Third, various views on the lessons are spread around two focused lessons – one from England and the other from China – in the wake of group discussions among teachers locally and internationally.

Seeing maths teaching through the researchers' eyes

The in-depth analyses of all lessons suggest that maths teaching tends to be similar within each country rather than across countries. There are typical lesson flows and key characteristics found across classrooms in each country. This is consistent with quantitative results on teaching. Such an intra-country similarity might be partly due to the influence of the uniform educational policies within each country. The following subsections will detail the researcher's interpretation/perspective of each country's maths teaching regarding its *typical lesson flow* and *key characteristics*. First, the English lessons will be illustrated, and then the Chinese lessons.

Typical lesson flow of the maths class in England

A typical English maths lesson starts with Success Criteria, is carried out through speedy first input and long repetitive re-inputs, and ends in Success Criteria.

Start with success criteria

After a pre-lesson practice of basic skills, such as times tables, English teachers typically start the lesson formally with a slide showing steps of *Success Criteria* after which they all name their lesson objectives. Teachers tend to be very open to share all lesson targets – big and small – with the children at the very beginning.

Speedy first input

Children are sitting on the carpet in front of the IWB while the teacher teaches the whole class step by step the lesson objectives. At the beginning, the teacher is interacting with the whole class, but this does not last long. The teacher uses this process to screen children off by ability judging by the quality of their responses or work on their mini WBs, and then from time to time the teacher sends off some of the pupils to the TA for extra support or back to their seats for the level of worksheets (often coloured according to the degree of difficulty) that the teacher thinks suit them. Those remaining on the carpet are able to progress to the next more advanced level of learning, and the ones that are last sent away are what teachers call the high-ability children. A typical maths lesson in England lasts about 50 minutes, but this selective input process only takes about 10 minutes.

Long repetitive re-inputs

By the time the high-ability children finally leave the carpet for seatwork, those low-ability children who have first been sent away are already back for extra input because they cannot cope with the tasks they have been given. Now, the cycle of "re-teaching – re-sending away – returning to the carpet – re-teaching" begins. Children at all levels keep coming back to the carpet for more support or for a new input to progress to the next (more advanced) step on the Success Criteria. Partial class interactions all happen at this stage. From the beginning of the lesson, the teacher's attention is gradually zooming in from the whole class to individuals, and those that later on fall out of the teacher's attention are supposed to be either doing independent work or self-assessing/marking their work with the pre-assigned answer sheet(s).

End with success criteria

During the last several minutes of a lesson, the teacher asks all pupils to gather on the carpet and reflect upon the Success Criteria steps; children then make their ways back to their seats, each self-mark their Success Criteria and stick it onto their workbooks (plain A4 notebooks). They are asked to bear in mind what they have achieved today and what they aim to achieve tomorrow. The lesson typically ends here.

Key characteristics of the maths lessons in England

With the typical lesson flow of a maths lesson in England in mind, we will now look at the eight typical features of an English maths lesson sequentially:

- differentiated teaching content,
- learning targets unfolded at one shot,
- direct transmission of knowledge,
- repetitive re-inputs,
- weak foundation and lack of readiness,
- lack of accuracy in teaching,
- frequent physical movement of pupils,
- support available for special educational needs.

Differentiated teaching content

English teachers spend more time in differentiating teaching and allocated more time for children working on ability-based worksheets. Though most schools are set by ability across the same age group, in-class grouping is also commonly practised in every classroom observed. Findings from unstructured observations coincide with results of a teacher questionnaire item (#38) which asks teachers to rate frequencies of a list of activities over a school year. These activities included *different work for different-ability children* which all 10 English teachers did "in almost every maths lesson".

Learning targets unfolded at one shot

English teachers show directly what children are expected to learn at the very beginning of each lesson. The embedded idea is to set a destination and then get there with the destination in mind. This sounds nice and seems transparent, but the teacher also loses the chance to nurture curiosity (what will turn up in the lesson) among pupils and make good use of such curiosity to deliver every chunk of knowledge at a right moment.

Direct transmission of knowledge

A typical approach to developing cognition in the English maths class is that the teacher demonstrates or states the standard way of solving a specific type of problems, with pupils following the way and practising the exact procedure during independent work. This approach eventually falls into the category of direct transmission of knowledge which leaves considerably little chance for gradual development of children's cognition in a coherent manner.

Repetitive re-inputs

Because of ineffective first input of each knowledge point, individual pupils keep coming back to the carpet for extra support, hence repetitive re-inputs from the teacher. If the first input did not work, the teacher should have asked her/himself what else she/he could do to teach the same content more effectively before her/his first attempt to re-teach.

Weak foundation and lack of readiness

Across classrooms, English pupils show certain knowledge gaps, such as times tables, place value, subtraction and addition, which should have been built solidly enough before the observed lesson. Such gaps from time to time prevent them from advancing to higher-level knowledge, which has not, however, drawn much attention from the teacher, let alone motivate the teacher to stop the next new input and start to consolidate the old knowledge. When there are pupils coming back for re-input, the teacher could have already thought of two possible reasons: (1) there is something wrong with the previous teaching method; (2) the children have knowledge holes that prevent them from moving towards more advanced knowledge. The teacher should be able to accurately predict the possible reasons that have prevented children from understanding what she/he had just taught and then take action to solve the problem immediately. For example in Teacher EN6's lesson on fractions, most pupils are not familiar with times tables – an essential foundation for learning fractions. Take the following interaction between the teacher and three children in the class as an example. To find 3/4 of 16, one needs to be proficient with 3 and 4 times tables which happened to be these pupils' weak point:

TEACHER: How many sections, Andy?

ANDY: 16.

TEACHER: 16 sections. OK. (writing down 16 on IWB) If I want to find 3 quarters, I need to find one quarter (writing down "1/4 of 16"). How many fours in 16? (looking at the boy next to her and waiting)
(. . .*15 seconds* later . . .)

ANDY: 4.

TEACHER: Good boy. 4. (writing down "4") How many quarters do I need? (looking at Amy)

AMY: 3.

TEACHER: So all I need to do . . . I need 3 lots of the answer. (pointing at "4") I've got my answer, and times it by . . .?

AMY: 3.

TEACHER: Great. (writing down "× 3 =") So, 4 times 3 is . . . ? (posing her fingers in a counting gesture and looking at Amy)
(. . .*27 seconds* later . . .)

4 times 3 (again with fingers in the counting gesture), 3, 6, 9, . . . ? (looking to Amy)

AMY: 12.

TEACHER: 12. (writing down "12" behind the "=")

Probably because of heavy reliance on the calculator from a very young age, English children show considerably weak calculation skills and number sense across classrooms. Teachers often suggest them to count fingers, which results in many children doing so in the process of completing tasks.

Lack of accuracy in teaching

Across classrooms, English teachers tend to write down inaccurate statements or answers and their handwriting also tends to be informal. For example, in Teacher EN6's lesson, to represent part of a whole, the teacher sometimes wrote it as 1/3 of 18 = 6, which still made sense though it should be formally written as $1/3 \times 18 = 6$, but sometimes the teacher simply omitted "of n", for example, writing $1/3 \times 12 = 4$ as $1/3 = 4$.

Similarly, when Teacher EN8 was trying to demonstrate 17% of 840 was 142.8, she made two mistakes: (1) an inverse percentage sign after the number 17 and (2) an incomplete expression which should be written as $17\% \times 840 = 142.8$ rather than $17\% = 142.8$.

Such inaccurate writing on the one hand may cause confusions among pupils in the specific lesson and, on the other, may also set up wrong examples for children to copy. If one day the teacher finally spotted the inaccurate or wrong expression in children's work, she/he would have no idea when, where and how the misconception had been formed.

There are also complete mistakes. For example, at the end of the lesson EN8, the teacher was explaining the answer to the final task. Everything went fine until it was time for the 6th column which demanded an increase of 5% from the previous answer 126 (Figure 7.1). The teacher took "+5%" as " – 5%", and none of the pupils found it wrong or, if they did, none of them pointed it out. Of course, the final result was absolutely wrong. If she had prepared the answers beforehand, she would have realised the difference, but apparently she had not.

Teacher's jotting & answer			7	12.6	6.3		
Start	-50%	+10	-10%	×2	+5%	+4	**Finish**
120	60	70	63	126	119.7	123.7	

A pupil's work

Start	-50%	+10	-10%	×2	+5%	+4	**Finish**
120	60	70	7	14	19	23	

Figure 7.1 The 'silent' mistakes in Teacher EN8's lesson

The bottom part of Figure 7.1 shows a pupil's work randomly captured by the camera. It was wrong from the fourth column (i.e. – 10%) onwards. Apparently, the pupil was not completely clear about the rule of the game, which had not been spotted by the teacher and probably would never be. When she wrote the final "answer" next to the last cell, the teacher announced that the lesson would stop here and "Table 2" would be dealt with sometime after the lesson.

Without clear handwriting and correct demonstration, any intended content would not be successfully understood and grasped by pupils. The results might be even worse – the lesson time had been wasted, more confusion had been embedded and new learning obstacles had formed.

Frequent physical movement of pupils

Activities happening in the English lessons often involve the change of pupils' seating/location in the classroom. In every classroom, pupils move from their seats to the front carpet at the beginning and end of the lesson for whole class interaction and during the middle of the lesson for multiple times mainly to receive re-inputs from the teacher either individually or in a group.

Support available for special educational needs

Apart from the classes EN3, EN4 and EN10, all other English classes had at least one teacher assistant (TA) present to offer extra support to children lagging behind or with special educational needs. In Teacher EN5's class, there were three TAs. This was different from the case of the Chinese classroom where only one adult – the teacher – was available for all.

Having seen the flow and characteristics of English lessons, we will next look at the flow and key features of Chinese lessons.

Typical lesson flow of the maths class in China

A typical Chinese maths lesson starts from and ends in real-life situations, and in between is a systematic problem-solving process in which the teacher plays a less active role and the pupils need to think and discuss actively the solutions to a sequence of tasks. Teachers seem to take the sequence seriously.

Start from real life

Chinese teachers tend not to tell what is in the bag at the start of a lesson. They all start from real-life issues or stories that are relevant to both pupils' life and the lesson content, pose open-ended questions and then seek the "right moment" to point out the title of the lesson. The intention is obvious – to draw pupils' attention without being too eager to load the burden of the whole-lesson's objectives onto pupils' minds at this stage.

For instance, to teach the lesson on place value, Teacher CN4 started his lesson from the way ancient people counted sheep, using stone substitutes and the idea of base 10. In a consolidating lesson on numbers from 10 thousands to 100 billions, Teacher CN9 started with a riddle about the sun and a discussion about the distance between the sun and the earth. Teacher CN8, to start her lesson on *Using Strategies to Solve Word Problems*, asked pupils to reflect on and talk about one of the Aesop's fables, *The Crow and the Pitcher*, which led to the key word of the lesson – "strategy".

The lesson by Teacher CN5 took place on 13 December 2012 which happened to be the 75th anniversary of the Nanjing (a.k.a. Nanking) Massacre. He started the lesson by posing a question to a pupil randomly, "Do you know what's the date today?" The pupil immediately responded, "The 13th of December." The teacher went on to ask the class, "Do you know anything special about today?" Almost everyone answered loudly, "It's the anniversary of Nanjing Massacre." Instead of going over this historical tragedy, he said, "Right, it's the 75th anniversary. Do you know the Museum of Nanjing Massacre near the Jiangdong Gate (江东门)?" "Yes!" answered the class. "There's a special number there. What is it?" asked the teacher. "Thirty *wan* (三十万: Chinese equivalent of 300000)," many responded immediately. Glancing around calmly but instantly, he asked, "What does the number mean, Ziyan?" The pupil stood up and answered, "It means that the Japanese killed thirty *wan* (300000) Chinese [in Nanjing]." Giving the class a nod, the teacher wrote 30 0000 on the board and said, "Thirty *wan*. Wow, there are so many zeros after 3. A very big number! Today we are going to learn very big numbers." Then he started to show pupils on the screen pictures about how ancient people represented numbers of objects using stones (one big stone representing 10 small stones etc.). The Q & A facilitated by the teacher then guided the class through a quick and vivid revision of place value of units through thousands, after which the new knowledge, four higher places next to thousands, gradually entered the scene.

Task-oriented process with systematic reasoning and problem-solving

The main part of a Chinese lesson is a gradually unfolding process in which the teacher promotes children to advance step by step through the lesson content. At the heart of the content are two to three exemplary problems that underpin the intended subject knowledge. The class only deals with one problem at a time, from simple to complex. Each exemplary task generally involves five steps: (1) the teacher poses the problem and pupils discuss it with their desk-mates quickly; (2) the teacher interacts with the whole class through intensive Q & A; (3) one or multiple solutions are generated by the whole class; (4) children try the solution(s) with a similar task and the teacher quickly circulates through the class to check everyone's work and collect representative work examples; (5) the teacher shows pupils' work examples – both correct and wrong – to the whole class via a projector and initiates whole class discussion regarding the answers. Steps (2) and (5) usually take longer time than other steps, which allows pupils to thoroughly practise reasoning and metacognitive thinking on the specific type of tasks, hence consolidating the newly obtained knowledge in time. Step (5) is particularly helpful in enriching pupils' learning experience with typical correct and wrong solutions carefully chosen by the teacher during seatwork. Step (5) also involves lots of reasoning on why certain solutions are right or wrong and how an alternative or wrong solution has been produced. Every single task thus demands sufficient time and thorough consolidation, and once it is done, there is no playback.

End in real life

Chinese lessons are all linked back to real-life situations to allow pupils to think about why and how the mathematical knowledge may relate to the real world and/or help solve problems in life. For example, to end lessons on higher-level numbers (10 thousands to 10 millions), Teachers CN4 and CN5 showed pictures about air pollution to the class and asked them to share their thoughts.

Key characteristics of the maths lessons in China

Chinese maths lessons carry eight main features which are as follows and will be introduced sequentially afterwards:

- pupils as active thinkers,
- wide existence of metacognition,
- solid foundation and absolute readiness,
- pupil classwork as teaching resources,
- optimal timing of teachers' writing on the board,
- teacher-guided and pupil-centred whole-class discussion,
- less physical movement of pupils,
- large class size without special educational support.

Pupils as active thinkers

Chinese teachers are more active in posing questions so as to generate solutions among pupils but *less* active in providing solutions. The duty of looking for an answer or solution to a given problem is on the pupils' shoulders. The degree of a Chinese teacher's input and support gradually decreases as she/he guides the class through the lesson. For instance, the first exemplary task in Teacher CN1's lesson on Data involved the collaboration between the teacher and the pupils; the following three tasks then saw a decrease of the teacher's input and an increase of pupils' independent effort; for the last task, the pupils worked in an absolutely independent fashion and completed it over a very short period of time – 33 seconds, all done!

In Chinese classrooms, the majority of the lesson time is spent on whole-class reasoning and discussion about questions as to how to solve the exemplary tasks, whether there are alternative methods, why certain emerging answers are correct or wrong and which step goes wrong if an answer/solution is incorrect. Pupils learn from each other's experiences throughout the lesson, and the teacher always sees the whole picture of pupils' knowledge development and their readiness for the next bit of teaching and learning. Moreover, in Chinese lessons, there is constant indirect reasoning about why mathematics should be learned, how mathematics is connected with the wider world and pupils' life and how knowledge obtained in the classroom might be applied in the real world. Children are motivated to learn actively all the time.

Wide existence of metacognition

One of the typical features in Chinese classrooms is the existence of lots of metacognitive discussion promoted by the teacher and carried out by the pupils particularly before and after each task. Before tackling a problem, pupils are asked to think what they must bear in mind while looking for a solution and why. This is often generated through intensive questioning and reasoning in the whole class. Upon task completion, stimulated by the teacher questioning, pupils think and talk about issues as to what the appropriate methods and correct results are, why certain mistakes have been made by themselves or other pupils and what measures should be taken to avoid similar mistakes in their future problem-solving.

For example, Teachers CN4, CN5 and CN7 all taught the same content – recognising whole-*wan*[1] numbers (认识整万数). While details about Teacher CN7's lesson may be found in teacher focus groups, we will here only highlight the similar feature that this lesson shares with all other lessons on the same topic. All lessons feature the teacher frequently asking pupils to talk about their methods of spelling and writing numbers. When being asked to describe the way she read the number 960 0000, a pupil in CN4 said, "I counted four zeros starting from the units, drew a Level-separating Line on the left side of thousands, firstly read 960 and then read the four zeros in the lower places as one word – *wan* (万)." As usual, the teacher followed with a question to the whole class, "Is her method right?" "Right!" responded the class immediately. When being asked to talk about the method of writing the number "nine hundred and sixty *wan*", another pupil said, "Firstly I wrote 960 and then I added four zeros behind it." The teacher quickly double-checked whether the method was right with the class, as if he did not know the answer at all.

Solid foundation and absolute readiness

Across all Chinese maths classes, children show their solid foundation and absolute readiness for the new content to come in. Chinese teachers seem to be more aware and confident of their pupils' level of knowledge and thus make more accurate 'prescriptions' for their capacity of learning the new content. If a teaching plan is a hypothesis, then Chinese teachers are good at making accurate ones, which is all down to their profound and thorough knowledge of mathematics and its teaching methods. Throughout the Chinese classrooms, the last task in each lesson is finished more quickly than the speeds for previous ones. In most cases, it is completed in less than a minute. As the lesson time goes by, pupils are becoming less dependent upon external guidance (teaching) in dealing with the intended content and more dependent upon their own newly acquired knowledge and experience (internal guidance). Sound foundation leads to absolute readiness. Effective teaching leads to a steadily growing body of solid knowledge base in pupils' minds. A positive cycle of effective teaching and learning is thus formed and constantly reinforced.

Pupil classwork as teaching material

A common phenomenon in every classroom is that during every independent work period (often 2–3 min per each), the teacher circulates very quickly through the class, checking everybody's work and identifying and picking up typical work samples (correct and wrong). She/he then shows the work samples through the projector and asks pupils to comment on them. Through an intensive round of whole-class Q & A, pupils' classwork is naturally turned into teaching material. During whole class interaction periods, pupils are often invited to come forward and demonstrate or explain their solutions. The use of pupil classwork in the whole class discussion enriches pupils' learning experiences through their observation and thinking of their own and others' successes and failures. For example, in his lesson on *Knowing Numbers at the Wans Level*,[2] Teacher CN5 showed the exemplary representations of 195 0000 and 200 0000[3] operated by pupils on the abacus.

Optimal timing of teachers' writing on the board (板书)

Chinese teachers all use both digital and physical (traditional) presentations as a way of representing mathematical content in their lessons. Of either approach, the typical feature is that they time 'accurately' (according to their own plans) when to write what on the board. The contents on the board by the end of a lesson would always be but not limited to the title of the lesson and the structure/key points of the lesson. Essential words will only be written on the board at the time point when the teacher feels key ideas have emerged naturally during the whole-class discussion.

For example, CN1 started her lesson on Data (sorting data by range) like this:

TEACHER CN1: Every Monday (a fixed day for the national flag rising ceremony in all schools nationwide) you will wear school uniforms, right? You look nice and tidy in uniforms. I would like to ask you what is the one thing that you should do before ordering your uniforms? [Pause three seconds] Haoxiang.

HAOXIANG: We should think and see which size of clothes suits which body size.

TEACHER CN1: Very well. [talking to the class] We should measure and see your heights, right? Okay, what if everybody has a different height? Shall we order a specific size of uniform for every one according to the height?

ALL PUPILS: No.

TEACHER CN1: Then, what shall we do? [pausing for 2 seconds] Luolan.
LUOLAN: Wear the size that is appropriate for one's height.

Then the teacher showed a paragraph of words on the first PowerPoint slide [left hand side of the board] which meant in English as follows:

> Meifeng Primary School is going to order uniforms for members of its drum and trumpet team. There are three different sizes. The small size suits those with heights of 120–129cm, the medium size 130–139cm, and the large size 140–149cm. Do you know the number of uniforms that the school needs to order for each size respectively?

She invited a girl to read the paragraph for the class, and as the pupil read, she wrote on the board just by the right edge of the screen vertically: "120–129", "130–139" and "140–149". When the girl finished reading, the teacher also completed her writing and turned back to the class and restated quickly the three ranges of heights.

Then, the teacher asked what else the school needed to know to book the uniforms and, after a pause of several seconds, she asked one of those who put their hands up to answer the question. The pupil said that, in addition to the size-height information provided in this paragraph, the school must also know the height of every member of the drum and trumpet team before making the order. At this point, the teacher turned to the next slide which showed 40 pupils' heights in a table. The three ranges (categories/sizes) of heights that the teacher wrote on the board were right beside this new slide showing the key information from the former slide.

Pupils were given three minutes to try and sort the data on their own. Then the teacher initiated a "three-party collaboration" between the teacher, a volunteer pupil and the rest of the class to categorise the 40 pupils' heights into the three ranges on the board using Chinese tally marks. The whole class Q & A went on and resulted in a more formal table based on the one with tally marks. These two tables were built on every answer the teacher received from the class, so the teacher acknowledged that these were entirely all pupils' work. Near the end of a conclusive Q & A in the whole class, the teacher asked the class again the number of ranges in the table. Immediately after the class said three ranges, she pointed out the lesson title, 分段统计 (Sorting Data by Range), and wrote it on the board at 00:18:58 in the lesson.

Likewise, after writing down the number 300000 at the beginning of his lesson, Teacher CN5 pointed out that this was a big number and today's lesson was about knowing big numbers. While he was saying the title, he turned to the blackboard and quickly wrote down the two Chinese characters: "认数 (English: Knowing Numbers)". In the whole class interaction, number places emerged one by one from the right to the left. The teacher added these places one at a time on the board and eventually formed the table consisting of four lower places (from units to thousands) which was old knowledge and four higher places (from *wans* (万， Chinese version of 10 thousands) to thousand *wans* (千万， Chinese version of millions)) which was new knowledge.

Similarly, 9 minutes 42 seconds into the lesson by Teacher CN6, he felt the moment was just right, so he pointed at the first round voting result of World Expo 2010 bids shown on the slide and said, "Today we are going to learn Bar Charts," writing the title of the lesson on the board, 条形统计图 (Bar Charts), as he spoke.

As Teacher CN6, all teachers tended to write notes on the board, while simultaneously speaking to the class. This dual (writing and speaking) emphasis of the lesson title and key points is a common behaviour observable in every Chinese classroom. Chinese teachers seem to use the screen (often PowerPoint slides) to show the flow of mathematical information and leave the chalkboard as a sacred space to showing the lesson title and key elements. The teacher will only write an element on the board when she/he finds the time is just right. Not only the teacher but also the pupils are well aware of the importance of notes on the board. The combination of the screen and the board shows the class both the dynamic (connections) and static (structures) characteristics of mathematics.

Teacher-guided and pupil-centred whole-class discussion

In a Chinese maths class, the teacher's role is more like a guide who ensures that the teaching and learning is heading in the right direction and who at the same time leaves the main job of mathematical reasoning and problem-solving to pupils themselves through asking lots of interconnected questions. It was through whole-class interaction and brainstorming that the guidance of the teacher and the central role of *all* pupils simultaneously come true. Teachers are also constantly encouraging pupils to open their minds, come up with multiple solutions to every single problem and explain them to the class. In this process, children have opportunities to think and tackle a given problem from different angles. The whole interactive process thus takes their thinking beyond the surface level of the problem and deepens their understanding of the intended mathematical knowledge.

Less physical movement of pupils

Contrary to the case of England, pupils rarely move from one place to another in any Chinese classrooms. Order seems particularly important in Chinese classrooms, and the collective effort between the teacher and pupils in keeping the order is an observable habit. There are several occasions in each Chinese classroom when individual pupils are asked to come forward to write answers/solutions on the board or manipulate instructional tools, such as the abacus, but only for the purpose of whole-class discussions.

Big class size without special educational support

The Chinese classroom has a bigger class size than the English classroom. Apart from Class CN7 which had 28 pupils at the time, the size of all other eight

Chinese classes ranged from 35 to 43. Although in China disabled children all attend special education schools, there might be some pupils attending state schools who at times also need extra support to meet their needs. The large class size seemed to work fine in all participating classes in the study, but the focus of the only teacher on the majority of children in one class meant several low-ability children might possibly easily be ignored in the teaching and learning process.

However, judging by the EMT data, we cannot see anyone in the Chinese classroom being significantly lagged behind and demanding extra support from a teacher assistant. Evidence in the literature suggests that more effective teachers are able to teach effectively in the whole class with a large class size (Good & Grouws, 1977). We thus hold our views upon whether there is a need to reduce the class size and put a TA in the Chinese maths class. Maybe it is because of the bigger class size that the learning experiences are richer; maybe it is because of one teacher teaching the whole class that the constant decision-making as she/he teaches is much easier to do in her/his own mind than having to discuss with another adult from time to time in a lesson of about 40 minutes.

Now that the researcher's unstructured observations of two countries' maths lessons have been presented, the following section will provide a picture of teachers' individual views on their teaching in general and observed lessons in specific.

Seeing maths teaching through individual teachers' eyes

Teacher beliefs of effective maths teaching

This section will sequentially present national pictures of (1) teacher beliefs of what counts as an effective maths lesson, (2) teacher strategies of organising a maths lesson and (3) teacher strategies of differentiating teaching.

What is an effective maths lesson?

RELEVANT TO REAL LIFE

Three English teachers (EN1, EN4, EN10) thought that an effective maths lesson should have a real-life meaning and children should be enabled to apply what she/he had learned. Teacher EN1 put it in this way:

> It's not just based in the classroom. They can think outside. And it's the application of it really. For example, decimals, they can apply it to money.... It's like connecting things up to apply them to the real-life situations.

However, as we have seen in the classrooms, English teachers appeared to teach pure mathematics rather than mathematical content related to or based on real-life issues/contexts.

DIFFERENTIATED AND CHILD-CENTRED

It is not surprising to hear differentiation and child-centred ideas from teachers in a system deeply influenced by its understanding of progressive education. In the English teachers' eyes, an effective maths lesson should:

- be appropriate to what the children had already known, be challenging enough to help them learn something new and be based in assessment of what children's targets were and the levels of curriculum they were at (Teacher EN2);
- enable all children to learn at the level that was right for them, make progress and think about what their next steps of learning were (Teachers EN4, EN8, EN10);
- appreciate that "not every single child is going to learn in the same way" (Teacher EN7);
- regard differentiation as the soul of an effective lesson (Teacher EN9);
- be appropriate for children's levels and involve self-assessment by the children themselves as well as the on-going assessment by the teacher (Teacher EN6);
- offer lots of opportunities for children to learn without involving too much teacher talk (Teacher EN3);
- engage children and move them along in their learning (Teacher EN5);
- enable *all* children to make progress – whether they were of high or low ability – and have input for every child in the class (Teacher EN10);
- not see behaviour as an issue; not tell children to do something; allow them to be involved in setting up targets and deciding where they start working or they are going to work on; set no ceiling to the children so that they are free to move (Teacher EN2).

AIMED FOR AND WELL ACCEPTED BY PUPILS

In Chinese teachers' mind, an effective maths lesson should:

- see pupils react to it well in the class, which should be the golden rule (Teacher CN2);
- be delivered in a specific method and easier for pupils to understand immediately and give opportunities to children for them to showcase their work, be their work correct or wrong (Teacher CN3);
- enable children to learn the knowledge *solidly* in a happy and relaxing mood (Teacher CN10);
- take both the environment and the specific learners into consideration, facilitate a very high level of active learning and consider the differences among pupils (Teacher CN1);
- "shine with the glory of humanity" and put children at the heart of it (Teacher CN8).

HAVING REAL-LIFE AND LONG-TERM MEANING

Five Chinese teachers also thought an effective lesson should:

- be helpful to the children throughout their life and add up to their ability in meeting challenges in their future work and life (Teacher CN4);
- have the premise that the teacher is crystal-clear about the context of the knowledge and the connections between the knowledge and children's life (Teacher CN3);
- enable children to synthesise knowledge (Teacher CN5) and apply the knowledge obtained in their life (Teachers CN3 and CN5);
- aim to get the pupils to grasp the methods of learning in the short-term and, in the long run, to build on the short-term goals to help them form good learning habits so that the children would still be able to apply the attained methods and habits in their future learning and continue progressing on their own (Teacher CN1).

WELL PLANNED

Two Chinese teachers also expressed that an effective maths lessons should be well prepared, carried out with certain flexibility (Teacher CN5) and based on the condition that the teacher knows clearly the key points and difficult points of the lesson and appropriate approaches to addressing the lesson content (Teacher CN3).

HAVING THE LESSON GOALS ACHIEVED

Four Chinese teachers were much concerned with the results of teaching and they thought an effective maths lesson should:

- see the intended results come true (Teacher CN6);
- ensure that lesson objectives have been achieved, that pupils have grasped what was taught and that their ability has been well developed (Teachers CN7 and CN9);
- equip pupils with the strategy of solving given problems (Teacher CN8).

Having heard teachers' answers to the first interview question – what counts as an effective maths lesson, we now move on to the following subsection which shows their daily strategies for organising maths lessons.

How do you organise and manage maths lessons?

ASSESSMENT-AIDED DIFFERENTIATION

English teachers all came up with this one and only approach to organising and managing maths lessons. Because of the typicality of differentiation procedures within schools, we present teachers' strategies school by school.

School EN-A

Teacher EN1 thought the ability-based set was helpful for the TA to come over and help a group if there were any maths misconceptions. He anticipated that it would be more than likely that someone of the similar ability had got that misconception as well, so then the TA could address the whole table. He stressed the importance of independent learning in maths, for example, providing pupils with an answer sheet so they could self-mark their work. Teacher EN2 generally started a new unit with a quiz to pre-assess where children were so that they could decide which step was appropriate to start on the Success Criteria. She would then normally focus on a group of pupils and assess whether they were aiming too high. In the end of a lesson, she would use a mini-challenge to assess how far children had made during the session.

School EN-B

Teacher EN3 clarified that his first strategy was to try to differentiate his class (the top maths set of the year group) in three ways: lower, core and upper activities. He often chose to be with the lower activity, which he said was due to the lack of a TA in his class. His second strategy was to do what he called "prior learning" at the start to see what children could and could not do. This helped him to decide who could do what. From the same school, Teacher EN4 also emphasised the importance of knowing the learning objectives and breaking them down to steps of learning in the Success Criteria. To know who could do what, she would generally do a prior activity to group children according to their levels. Also without a TA, she expressed the difficulty of meeting every child's needs at different levels as she planned. Her follow-up strategy was:

> What I do is I do a record of where they are at the end of each lesson so that I know the next day what learning they need or what group they need to be in.

School EN-C

Teacher EN6 usually started with a "mental starter" to get the pupils thinking and then moved on to the main activity, with separate inputs serving for children of "a wide-range of ability". She generally assessed children's WBs on the carpet and then sent some away to independent work or brought them back for the next step of learning. Teacher EN7's strategy was to assess children's levels at the beginning of each week so that they could be grouped by ability:

> If the children get every single one right, they'll be on the top table; if some get a few of them right, they'll be on the middle table; if none of them, they'll be on the bottom table.

She would then plan a week of lessons to move the pupils' learning on. Having said that, she pointed out that her plan was flexible and everything depended on how pupils got on and how she taught them. "So I always change," she said. Teacher EN8's strategies were to assess children at the beginning of the lesson to have an idea "where children already are". She would then differentiate her teaching accordingly. Teacher EN9 also saw differentiation as the core of her strategy, though she pointed out that it was important to be flexible at the same time. Teacher EN10's main strategy was differentiation through the "cutaway style" as called by teachers at the school. With this method, the teacher grouped children by assessing and sending some of them away for independent work at each step at the start of a lesson. This often happened on the carpet where children's work on their mini WBs would be swiftly viewed and assessed by the teacher.

Chinese teachers offered a wider range of strategies than did their English colleagues. These strategies include (1) paired or group discussion, (2) the use of warm reminder, jokes or riddles, (3) matching the characteristics of each lesson, (4) promoting peer instruction among pupils and (5) being well prepared and flexible.

PAIRED OR GROUP DISCUSSION

Chinese teachers talked about a variety of ways to organise and manage their maths class. In terms of grouping, their aims were not about streaming or setting but about putting different-ability children together in a group and optimising every child's learning through offering rich opportunities for them to learn from or be motivated by each other. They all considered that grouping varied depending on the purposes of different tasks in each specific lesson. Overall, their approaches included paired discussion for normal tasks (Teachers CN2, CN3, CN4, CN5, CN6), collaboration in groups of four for more practical/explorative tasks (Teachers CN3, CN4).

THE USE OF WARM REMINDER, JOKES OR RIDDLES

About managing class, three Chinese teachers also thought of and talked about ways of managing pupil behaviours and keeping everyone on task. Teacher CN3 would generally give a warm reminder to an off-task pupil but if there were *lots* of off-task pupils in the class she would reflect upon her own teaching. Teacher CN4 would make the lesson more interesting and less dull, through telling jokes and stories and using real-life situations. If there were bad behaviours, he would generally send information through eye contact or remind the pupil in a humorous tone. Teacher CN9 would use a game or maths riddle to boost pupils' curiosity, so that they would wonder why the teacher did so and what might be happening in the lesson.

The strategies that they talked about echoed with what was observed in the Chinese classrooms. What the teachers emphasised was the importance of observing and 'reading' children's reaction and emotion instantly, deeply and accurately and then immediately preventing off-task or behavioural issues from happening or continuing through less apparent teacher behaviours. Chinese teachers appear to have more reflection in action – one of the three types of teacher reflection that Donald Schön (1983) wrote explicitly about in his work.

MATCHING THE CHARACTERISTIC OF EACH LESSON

One Chinese teacher (CN7) mentioned that she tended to use different strategies in different lessons. In a calculation-oriented lesson, she would mainly ask pupils to do practice; in a lesson at the end of a unit, she would arrange more games and activities; in a practical lesson, she would ask pupils to interact with their desk-mates or in a group of four on a more complex task.

PROMOTING PEER INSTRUCTION AMONG PUPILS

During the Q & A process, if a pupil gave a wrong answer, Teacher CN7 would paraphrase her question and try to guide the pupil to get the correct answer independently; if it still did not work, she would ask others to help the pupil, rather than explain the correct answer directly herself. She insisted that pupils, instead of the teacher, should find the answer. For the development of those pupils who did not grasp the knowledge after thorough instruction, Teacher CN9 would show their work to the class via the projector during the whole-class discussion and ask other pupils to help them identify reasons for their mistakes and offer solutions.

BEING WELL PREPARED AND FLEXIBLE

Two Chinese teachers considered a great deal about things that they thought crucial at the planning stage of a lesson. Teacher CN8 said she would generally make a very detailed plan and accurate predictions of the lesson, therefore coping freely with what emerged in the lesson. Having said that, she thought the strategy should depend on the characteristics of each lesson. Teacher CN9 usually took time to design the questions in a way that every pupil could come up with a few ideas to share with the class in the lesson.

Overall, English teachers expressed a same strategy in organising and managing their maths lessons – differentiation – a concept that lies deeply in the heart and soul of progressive education advocated, interpreted and implemented in the West. Chinese teachers talked about more approaches and purposes behind them. It appears that Chinese teachers are more likely to be open-minded, foresee the effects of different ways of teaching, optimise the lesson plan and adjust their plan and action in the dynamics of actual teaching and learning.

How do you differentiate teaching in the maths class?

English teachers talked about four different strategies for differentiation including (1) self-differentiation, (2) pre-assessment, (3) support from TAs and (4) different abilities different tasks.

SELF-DIFFERENTIATION

Two teachers (EN1 and EN2) from the same English school (EN-A) both raised the idea of self-differentiation which was easily observable in their lessons. Considering the fact that the school did not have between-class setting (i.e. ability grouping), they believed that the steps in Success Criteria (lesson objectives) provided the opportunity for children to move through according to their self-judgement of their levels.

PRE-ASSESSMENT

Instead of asking children to self-assess, four other English teachers (EN3, EN4, EN6 and EN7) shared their way of classifying children's ability in each lesson through a certain form of prior learning activity where children's work on worksheets or mini WBs would be assessed. In fact, all English teachers but EN2 used this pre-assessment approach in their observed lessons. Teachers EN3 and EN4 said they did so by group. They would check pupils' answers by quickly looking at their work and immediately see which group needed more challenge, which group needed to "be brought in support", which group should be kept for a bit longer, and which group they should be working with. Teachers EN6 and EN7 shared their strategies of assessing and putting children into different groups. Teacher EN6 mentioned a maths app that helped her break down "each strand of maths into the levels" and gave her a rough idea of children at different stages. At the beginning of a lesson, she would quickly check children's WBs or reflect on their previous work and decide where they needed to go. Teacher EN7 continued with her weekly strategy in managing the class to detail her differentiation approach. She reemphasised that the beginning activity set at the start of a week helped her decide which group the children fell in. Then, the activity on the carpet in each lesson further helped her differentiate the kind of support she gave to different children. Having said that, she still found it "quite hard to differentiate".

The above two forms of differentiation-driven assessment – either carried out by the teacher or the pupils – are clearly based on the theoretical assumption of progressive education that learning goes through different stages. English teachers appeared to perceive the best way dealing with differences amongst children must be separating them from each other so that within groups they are at the same stage. It thus seems that both approaches fall into the same philosophical orientation. So do the following two strategies.

SUPPORT FROM THE TAS

English teachers all had at least one TA in place except Teachers EN3 and EN4 who uniformly pointed out the importance of TAs in assisting them achieve differentiation and both felt deeply that the lack of a TA considerably limited the quality of differentiation they would ideally want. Teaching a lower set with a TA, Teacher EN9 expressed her reliance on her TA to help meet different needs. As a colleague of Teachers EN3 and EN4 from the same school, Teacher EN5 had two – sometimes three – TAs for her bottom set class, but she still found there was always a group who could not "remember how to do it". Differentiation seems to be a heavy load for every English teacher with or without a TA.

DIFFERENT ABILITIES DIFFERENT TASKS

It is observable that in every English classroom children were given different tasks for apparent reasons – they were of different "abilities". Only three maths teachers from England, however, talked explicitly about different tasks for different-ability pupils as a method of differentiate teaching. Teacher EN8 found within her middle-set class "still a big gap" among pupils ranging from low level 3 to high level 5. Her differentiation was mainly between the sheets that children were doing and the amount of support that she gave to them:

> Do they need a quick chat with me? And move away? If they don't understand it, is that to break it down? Let's do an example on the board, and try one on your book. Still don't understand [it]. Let's do another on the board. Try one on your book. See how they go with it. I mean, there's a lot of time – some of the children on the carpet for a long time – but they're still doing their work.

Teacher EN9's strategy was to differentiate her expectations to different-ability children, expecting high-ability ones to use "a variety of skills and strategies" and those with low ability to use probably only one strategy. Teacher EN10 would also give different objectives – easier or more difficult examples – to the pupils in the top set (class), so that "they were all doing the same work but at different levels" that suited them.

Chinese teachers talked about quite different strategies for differentiation from the English approaches. Their strategies were (1) aiming for the majority, (2) questioning techniques, (3) post-lesson support to the low ability and special awards to the high ability, (4) put the high- and low-abilities in pairs or groups and (5) low-ability pupils to preview lesson contents.

AIMING FOR THE MAJORITY

All Chinese teachers mentioned that their lessons were generally tailored for the majority of pupils. Some pupils already got the knowledge before being taught in the class, so the teacher would encourage these high-ability children to speak up more to set examples for mid- and low-ability pupils. For those very few low-ability pupils, extra support would be provided after class.

QUESTIONING TECHNIQUES

Chinese teachers all described similar techniques of questioning: (1) for normal questions, children would be asked randomly; (2) for simple ones, low-ability children would be asked, so that they could feel they were at least capable of *some*thing, that they were more likely to stay on task, and that their self-esteem could be uplifted; (3) for difficult questions, high-ability pupils would be asked. For example, Teacher CN1 talked about her strategy in dealing with difficult questions. After a high-ability child correctly answered the difficult question, she would first ask those slightly lower than the high-ability pupils to re-explain the process of solving the problem, second ask the whole class to explain the solution to each other in pairs, and finally ask a low-ability child to talk about the solution again to the class. Thus, the teacher was able to consolidate the same subject content amongst a variety of learners, through asking different pupils different levels of questions and asking different pupils to answer a same question at different points in time. Teacher CN9 also said she would first ask high-ability pupils to do a challenging task and then invite middle- and low-ability children to re-explain the solutions so that the latter two levels of children could "absorb and transform the knowledge into their own" through the re-explanation.

POST-LESSON SUPPORT TO THE LOW ABILITY AND SPECIAL AWARDS TO THE HIGH ABILITY

The ability range in Chinese classrooms is quite narrow which is confirmed by the tests results and teachers' ratings in the questionnaire. For the few lowest ability pupils, all teachers' strategy was to mark and explain their homework to them in person. For example, after the lesson, Teacher CN9 generally asked other pupils to help the low ability, or she would help them herself if they still had some problems with their work. For those high-ability pupils, if they achieved 98 or 100% in any unit tests or mid-term tests, Teacher CN3 said she would issue them a "gold" medal which guaranteed a homework-exemption opportunity. The teacher said, however, none of them had used the medal once this term (i.e. September 2012 to January 2013), by which she seemingly meant those pupils actually very much wanted to do homework. In some of her lessons, the high-ability children might be given challenging tasks, or be authorised to read their favourite books while others were doing exercises.

PUTTING THE HIGH- AND LOW-ABILITIES IN PAIRS OR GROUPS

In some classes, subject teachers of Chinese, maths and English would discuss together as to who should sit next to whom according to their ability levels. They sought to put high- and low-ability pupils in pairs, so that the high-ability pupils could help their low-ability buddies. Teacher CN7 said her strategy was to organise pupils' seats through grouping them by four, with each group consisting of a high-ability pupil (#1), two middle-ability ones (#2 and #3) and a low-ability (#4). She would give specific questions during the lesson – challenging questions first for #1 to answer, then #2 and #3 to give supplementary answers and finally for #4 to conclude the answers. Conversely, about simple questions, the teacher would ask #4 to talk first, #3 and #2 to follow up second and finally #1 to give complementary/alternative answers.

LOW-ABILITY PUPILS TO PREVIEW LESSON CONTENT

Two teachers from School CN-D, Teachers CN8 and CN9, both mentioned their strategy of suggesting low-ability children to preview lesson content. During the lesson, Teacher CN9 said she would also give more opportunities for low-ability children to answer questions and respond to them with more positive feedback to keep them confident.

Summary on teacher beliefs

This section focuses on three strands of teachers' beliefs: (1) a description of an effective maths lesson, (2) the strategies for organising and managing maths lessons and (3) the strategies for differentiation in maths lessons. Overall, English teachers tend to have a narrower range of answers for all three questions than Chinese teachers; English teachers are also more concerned with differentiation and the utilisation of assessment for differentiation than their Chinese counterparts who focus on the whole class and seek to use questioning and methods other than assessment to facilitate the progress of different-ability children; Chinese teachers demonstrate deeper thinking than English teachers especially when talking about the above second and third strands of teacher beliefs.

The following section presents teachers' views and comments on their own lessons while and after watching the lesson videos.

Teacher self-evaluations of their maths teaching

The second part of the teacher interview consists of three questions regarding teachers' (1) reflection on the teaching plan, (2) self-evaluation of the implementation of the plan and (3) self-comments on the lesson and its effectiveness.

In Teacher EN1's plan, the first thing was to ask pupils to write the date on their workbooks and know what they would be doing. Then he would ask them questions about their marking on their workbooks the day before the lesson. Following that was a quiz that would assess children and tell them when (i.e. at the point they got stuck) they should leave the carpet and work independently on certain steps (i.e. differentiated worksheets). The teacher would then work with the group that needed more support. About the actual implementation, he thought the plan went as he expected. Having revisited his lesson through watching the video, the teacher justified a few things he did and thought the lesson was effective since children retained their times tables as he planned. Those things included (a) the use of steps of Success Criteria which told children what to do at the beginning of the lesson, (b) asking pupils to talk about reasons for learning times tables, (c) using the quiz as a screening/differentiating tool in which those who encountered difficulty at any point could leave the carpet and go to the TA's table for support and (d) opportunities for children to self-assess and self-plan next day's targets.

Also from the school EN-A, Teacher EN2 planned the unit – times tables – as a whole and the lesson observed was amid the unit. This unit had been taught and learned last year but this year it was re-taught and relearned. About how the plan went, she said it was certainly what had been planned to happen. Watching the Q & A at the beginning of the lesson video, the teacher was pleased to see children "very keen to share ideas" on why they should learn times tables. While other children were about to practise and self-assess times tables, several pupils were asked to join the teacher on the carpet. The teacher said that they needed extra support with their number bonds. Overall, she thought that the pupils were all on task, timing and challenging themselves, making progress and knowing where they could get things and how to move on, so the lesson was effective.

From School EN-B, Teacher EN3 planned to do a 10-minute prior learning activity after a "starter" (times tables practice) and then a 30-minute main activity on the calculation of perimeters and areas of different shapes and "ideally" bring pupils together in the last 10 minutes to allow them to self-assess their Success Criteria, stick it in their workbooks and put them into three colour-coded trays (green = I'm confident, yellow = I'm getting there, red = I'm not confident). The teacher thought that he had carried out his plan and realised the lesson objectives. The only thing that surprised him was that pupils were struggling with perimeters of irregular shapes (L-shapes) rather than – what he predicted – the areas, presumably because the jump from regular shapes (rectangles) to irregular ones was a bit too much for some pupils. A little regret was that the ending was a bit squashed. About the effectiveness of the lesson, he thought the interaction, paired discussion, paces and "all the elements" were there, so the lesson was effective, but maybe he had pitched it a little bit higher.

Also from School EN-B, Teacher EN4 planned to start with *a multiple of ten times by a unit number* and then *a multiple of ten times by a multiple of ten*. "The class has done this before so it would be a recap," she said. She thus hoped the children would be quicker than they were previously, so she had not planned to do a prior activity to assess where they were. She felt the lesson did not go as she planned, because there were "lots of misconceptions" among pupils. However, she said, "Other than that, all sort of went how I planned it." Throughout the interview, the teacher had not pointed out the lesson content which was in fact using the grid method to calculate the product of two two-digit numbers. Whilst watching the lesson video, she said that her middle set was a hard maths group where there were "a lot of behavioural children". Looking at the boy A raising his hand eager to talk, she said A is quite difficult, B is quite difficult, C, D and E – they were quite difficult. The main activity went "longer than usual". Overall, she thought the lesson was effective because the pupils had recapped the grid method.

Another teacher from School EN-B, Teacher EN5 originally planned to start with a times tables work and then a quick activity to practise a skill from yesterday – rounding up or down decimals to put it into money rather than just random decimals. The main part would be adding and subtracting decimals. The ending would be children self-assessing how they felt they had done, what they thought they needed to do and why they said they were just "yellow" (getting there) rather than "green" (confident). The teacher thought her plan "went OK", with lesson goals achieved. Nevertheless, she mentioned again about the difficulty in accurately predicting the speed in which this lower maths set (class) might be progressing in a lesson:

> you always thought, "I wish I got time to get that bit in", because you're never sure how long it's gonna take to recap things, you never know how quickly they gonna remember things they know. What you hope doesn't always happen.

Teacher EN5 said her pupils were good at staying on task and the lesson was more of a recap, but she did not give a clear comment on its effectiveness.

The following five English teachers were from School EN-C. According to Teacher EN6, her class had done a lesson, four days before the observed one, on the same content – fractions of a shape. In that lesson, the low-ability children were expected to work out 1/2 and 1/4 of a shape, the middle ones tried to figure out 1/5 and 1/6 of a shape and the top ones were asked to find 2/3 and 4/5 of a shape where the numerator was more than one. Four days later, the teacher planned to do a quick recap and then immediately move on to "trickier" factions of a shape where, for example, the sides or segments were different. The teacher planned to do more complicated fractions of a shape but soon found that the majority of the class could not remember what she had taught them four days ago. Then, the planned lesson became a recap:

> They found it trickier than I thought they would. They needed a lot more consolidation than actually, [judging] from the books that they showed me four days ago. They had got it but they got it with me. They needed now to practise it independently. So I had to spend a lot more time on that than I'd first imagined. But, that's just adapting the lesson to meet the children's needs. I found the low ability they moved on to the tricky one – the tricky fraction of a shape.

Watching the lesson video, she explained she sent both high- and low-ability children to the TA for more advanced learning or extra support. From time to time, she tried and talked to the TA to see whether everything went well, because the TA was not a qualified teacher after all. She highlighted the "traffic light" activity at the end of the lesson where children coloured the steps they had reached (green = I got it, yellow = a bit more help, red = I didn't get it). To conclude, she thought the lesson was effective.

In the lesson prior to the observed one, Teacher EN7 noticed that a group of children did not get to the point she wanted them to, so she planned to send this group to the TA in this lesson. She would work with the "main" group on the carpet about ordering decimals and factions. Looking back, the teacher still found some of the majority that she kept on the carpet for the new lesson did not get it, but she was determined to stay optimistic and flexible:

> You (the pupils) don't get it. Let's do some more on the carpet. When I actually sent my top table off, I noticed that they've done exactly the same wrong and that was my fault, cos I didn't say it. So I have to bring those children back to the carpet, say, "This is where we got wrong yesterday. It's my fault, I didn't tell you." I went completely off plan. They got everything, but not in the order. They met everything I wanted them to but it wasn't as quick and effective as it should've been, because I have to go over things again today.

She said that all pupils were on task and well-motivated, and overall, the lesson was effective because she had spotted what misconceptions the children had.

What Teacher EN8 aimed to do was to teach the low- to middle-ability children to calculate 50% of a given number and high-ability ones first to figure out 1% as well as any percentage of a given number, second to take a deduction off a number and finally to solve word problems involving these calculations. In reality, the content Teacher EN8 originally aimed to deliver was not fully covered – high-ability children did not get to the point of solving word problems. The teacher thought that the start was good, because children were successfully split into groups of calculating 50%, 10% or 1% of given numbers. Then, the even lower children had spent time working on dividing a number by 10 and dividing it by 100, which the teacher thought was going well, too. However, she pointed out that on the one hand she might have expected high-ability ones a bit too high and on the other she anticipated low-ability ones would need "a lot of explanation" but they actually moved on to the next level. Overall, the teacher said the lesson was effective because the children were appropriately differentiated and they all had the enjoyment of mathematics.

Having taught the children about fractions in the prior lesson, Teacher EN9 planned to teach decimals. First thing she planned to do was to use sectioned shapes to show children that a decimal, like a fraction, was also a part of a whole. Second, she planned to do decimal number lines to help children recognise the sequence of decimals, such as 0.1, 0.2 and 0.3. Third, the focus would be on place value of decimals and the understanding would be reinforced through labelling places of the number "23.4". The teacher thought the concept of place value was the foundation of learning decimals. In this part, she planned to do a card game in which children would manipulate their digital cards to represent the decimal numbers that the teacher read. The final step would be about ordering decimals.

The teacher thought she had stuck to her plan. She seemed particularly surprised to see the low-ability pupil actually progressed very quickly and had a much more solid understanding of what a decimal was. Thus to "adapt things slightly", she tried to get the child working more independently. She also found that lots of children understood quite quickly. After they could order sequences of numbers in one decimal place, they quickly figured out how to order numbers to two decimal places. She was delighted to see children progressing so quickly and having a "secure understanding of what a decimal was". Having seen her lesson video, the teacher said it was good to spend longer time – "longer than usual" – on the carpet for the input without which the children would not have the "secure understanding" that helped them progress and speed through the learning. Overall, she said the lesson was effective because every child, by the end of the lesson, could tell what a decimal was in terms of place value.

Teaching the only top set in the school, Teacher EN10 planned to build on yesterday's lesson about "percentage of a number and percentage off" to teach the calculation of increasing/decreasing a number by a given percentage. He planned to use football as an example and anticipated it would work catching children's attention. Reflecting upon the lesson, he thought all went as he planned, since every child in the class made her/his own progress and knew how to increase and decrease a number by a certain percentage. Though the level of the second task was a bit too high, but it was OK, he concluded. He appreciated their level of maths and thought some of his pupils were much faster than him. At the beginning, he regretted and said that he should have asked the pupils to talk to the person next to them. He pointed at two children and said they were quite good at maths – like mathematicians. At 00:10:20 on the video timeline, he said that he should have done some calculation instead of talking. He further mentioned that he usually prepared extra exercises for high-ability children. Overall, the lesson was regarded as effective because all children had made progress.

Based in School CN-A, Teachers CN1, CN2 and CN3 all planned to teach how to order and put a large amount of data into a table – the first session on Chapter 9: Statistics and Probability in the textbook, Mathematics 4A. Maths teachers in the same year group of the school always planned lessons together, maintained the same teaching pace and did the same assessment since they were using the same set of textbooks (provincially unified). In addition, they also assigned the same homework every day. Because they shared the same office, it was very common they discussed the results of children's homework while they were marking it.

All these three teachers thought that they had carried out what they each planned and that their lessons were effective, but they also each found small things that they regretted. Along with the process of demonstrating a table, Teacher CN1 thought she should have presented corresponding notes on the board at a right time accordingly. "Because we should leave on the board a complete set of 板书 (notes on the chalkboard) at the end of each lesson", said the teacher. Teacher CN2 regretted that she forgot to remind the pupils that in the final exercise there was not a table for ordering data like the one in the exemplary task. If she did, pupils would all know the only table provided must be filled in with exact numbers rather than Chinese tally marks "正". She also said that the lesson would have been better if she left more time for relating data to environmental protection. Teacher CN3 regretted that the middle part of the lesson had not been quicker, but she was quite happy that she had showed pupils' work via the projector to the class as a conclusion to every task. Teacher CN1 also highlighted a specific event that she thought went crucially well. It was what she called a "three-party collaboration":

> This process involves me writing on the board, a boy pointing at the data and the class telling me where I should jot down a stroke [of Chinese character 正]. I did this because I was still not very sure if everyone had grasped the method solidly. I had to consolidate at this point, because teaching examples is the most important part of a lesson. I must make sure every pupil clearly understand the method through the teaching and learning of ***one*** example. I can't patch up here and there later when misunderstandings are already there. This process gave every pupil a second chance to judge which entry went into which category. After this, they should have a clearer understanding of this method.

From School CN-B, Teacher CN4 planned to teach Recognising Whole-*wan* Numbers (认识整万数, see notes 2 and 3). The lesson goals were pupils should be able to read and write numbers at the *wans* level (万级, *ibid.*) accurately and quickly. He thought the reading and writing methods of numbers at the *wans* level was well structured, but the Level-separating Line[4] was not thoroughly taught. Overall, he said the lesson was effective since the lesson goals were realised.

A colleague from the same school, Teacher CN5 prepared the same content as Teacher CN4's, which was what they had usually been doing in the same year group. In planning the lesson, he also referred to the teaching guidance published by the provincial textbook publisher and aimed to help teachers plan lessons. Saying that the plan went well, however, he felt that there was sufficient teacher-pupil interaction but insufficient pupil-pupil interaction and that high-ability children's needs might not be thoroughly addressed. Overall, he thought it was effective since all his targets had been achieved.

Teacher CN6 said he usually maintained the same teaching pace as that of Teachers CN4 and CN5, but the lesson observed was about a different topic – constructing bar charts – the second session (pp. 75–78) in Chapter 9 Statistics and Probability. During the lesson planning process, he tried to build the lesson steps on children's prior knowledge to facilitate "positive transfer (正迁移)" as the teacher put it. They had learned bar charts in which a unit of the Y-axis meant one, instead of five or 10. This lesson was about constructing the bar chart where the unit of the Y-axis was a multiple of five, so the teacher planned to start from the old content and lead pupils gradually to the new. The teacher thought the lesson went well but not as perfectly as he imagined. Having said that, he regarded it as an effective lesson.

The only Grade-4 teacher from School CN-C, Teacher CN7 planned to start from the historical story about ancient people representing numbers through tying knots on a string/rope which would then remind the pupils of decimal system that they had already known. Then, the class would learn to generate the fact that when the number of beads in the thousands' column on an abacus reached 10, they should put the 10 beads back and add a bead in the place higher than the thousands. At that point, the teacher would introduce the *wans* place (万位, see note 2). After that, the other three digits at the *wans* level (万级, *ibid.*) would emerge and be introduced sequentially in the process of the whole class manipulating their abacuses and counting numbers. During the process, there would be reading and writing tasks. In the end, the teacher planned to present real-life pictures which involved numbers at the *wans* level, such as the picture of the Olympic Stadium in Nanjing. The teacher thought the lesson went well, except for the fact that she wrote the name of the thousand *wans* place at a later time than she planned. She also regretted having forgotten to introduce the Western numeral system, which separates numbers by three digits, and ask pupils to compare that with the Chinese numeral system which separates numbers by four digits. She also found the ending was a bit rushed, but she was satisfied with the fact that the lesson put sufficient emphasis on place value and the Level-separating Line. The lesson in her view was effective.

From School CN-D, Teacher CN8 planned to teach the unit on Problem-solving Strategies which on this occasion was about using tables/arrows to sort known and unknown numbers and then to solve word problems. Because it was the first time at the primary stage for pupils to learn "strategies" formally in the maths class, the teacher planned to start from the *Crow and the Pitcher* – an Aesop's fable that pupils had previously learned in their Chinese class. The teacher planned to ask pupils to tell the story and then to think about how the crow finally managed to drink the water. It was at this point when the concept "strategy" would be formally introduced. She would then pointed out that, in mathematics, people would also need "strategies" to solve problems. After that, she would show a picture of the class's weekly timetable and ask pupils what lesson they would have during a specific slot on a specific day. Next, she would ask why they could so easily find the answer, which would lead to the idea of using tables to get information organised. Following that would be the main part of the lesson. The intended strategy would be taught through two exemplary tasks. The key point was to get children to realise the importance of distinguishing useful information from useless information when there was more information than needed.

Looking back, she detailed several places where she should have done better. First, she thought she should have shown a very messy list of subjects scheduled for a week before showing the class their timetable, thus to form a bigger contrast between messy and well-organised information and allow children to see the beauty of organising information neatly into tables. Second, she regretted that, in the second example, it would be better if she let the pupils figure it out themselves rather than she suggested pupils to use arrows instead of tables to illustrate relationships between known and unknown numbers. She attributed the reason to the fact that time was limited. Third, she thought more emphasis should have been put on pupils' understanding of different quantities in a word problem and the relationships between them. Nonetheless, she said that a perfect lesson was not necessarily a good thing and that she had reflected upon her teaching as always and planned to make it up in the next lesson. Reflecting upon the positive side, she thought it was good to use the story of the *Crow and the Pitcher* at the beginning of the lesson which gave the children a vivid sense of what the word "strategy" meant. Like all other teachers, she considered the lesson effective.

Teacher CN9 was from School CN-D as well. She planned to review the lesson content prior to the observed lesson which was about whole-*wan* and whole-*yi* numbers (整万数和整亿数[5]). In the revision, she would ask pupils to talk about place value of some given numbers. Then, the main activity would be about numbers that consisted of non-zero numbers in each place at the *wans* and *yis* levels (万级和亿级 notes 2 and 5) and the lower places did not necessarily contain zeros. The end of the lesson would be a conclusion on what had been taught and learned. She said the lesson was carried out as she expected, but there were quite a few points that she thought could be better. For example, the beginning of the lesson was about the distance between the sun and the earth, and the type of that number was what the lesson intended to address. She thought it was an inappropriate example which would be better to appear later, but luckily pupils still managed to read it correctly. Another thing was that she should have stressed more on reading and writing numbers at the *yis* (亿) level (note 5). Nonetheless, she said it was good to stress the concept of the Level-separating Line. Overall, she also thought the lesson was effective with all lesson objectives achieved.

Summary on teacher self-evaluations

All with a clear vision as to what to do in the lesson, almost all teachers thought they had carried out the lessons as planned. At the same time, there were also some teachers from both countries regretting things that they had not anticipated or planned well, but the types of things on the regretting list were quite different. English teachers were likely to find out the difficulty level of content and/or the pace of teaching they foresaw did not match the reality, which explains why they had to give up plans in a varying degree to adapt to the reality. On the contrary, Chinese teachers who regretted were all happy with the kinds of things that English teachers mourned for, and what they actually regretted was the imperfectness of certain small steps.

Regarding the quality of the lesson, almost all teachers agreed that their lessons were effective *overall*. Two English teachers, EN1 and EN2, both from School EN-A did not find any particular elements of the lessons that they were not satisfied with, whereas the rest of the English teachers and all Chinese teachers each expressed a varying number of things that they thought were not arranged well enough. The majority of teachers showed critical thinking upon their own practices, which is an essential step towards improvement. Besides, Chinese teachers were more concerned with their ways of dealing with specific points of knowledge, while English teachers cared more about their approaches to grouping children and differentiating teaching and learning content.

International awareness and flexibility to change

In the last part of the interview, teachers were each first asked to describe their imagination or impression of the mathematics teaching to the same age group in the other country – in England for Chinese teachers or in China for English teachers. Second, they were asked to talk about their flexibility to change in their future practices.

With regard to a maths lesson to the same age group in China, English teachers thought it might:

- have desks set in lines, pupils facing forwards, one or two children per desk and the teacher delivering the lesson at the front (all English teachers);
- deliver more advanced mathematical content (Teacher EN1);
- have more able, self-driven and independent learners and allow less paired discussion (Teachers EN3, EN4 and EN8);
- have more individual learning, self-support and skills-based learning rather than the application of knowledge (Teacher EN1);
- be much more formalised and structured (Teachers EN2, EN3 and EN4), more adult-led than child-led (Teachers EN2, EN7 and EN8) and more about rote learning and a lot stricter (Teacher EN5);
- have everyone on task (Teacher EN5);
- use whole class teaching and textbooks (Teachers EN6 and EN7);
- offer fewer opportunities to children to come to the teacher and talk on the carpet (Teacher EN8);
- have little differentiation and challenge and care less about individual needs (Teacher EN9).

In the Chinese teachers' imaginations, an English maths lesson would:

- have no more than 20 children with several teachers who would give them elaborate attention and care (Teachers CN1 and CN4);
- have easy curriculum, slow pace, low expectations and loose standards (Teachers CN1, CN7 and CN8);
- be more flexible and more related to children's life experience (Teachers CN7 and CN9);
- have pupils sitting at round tables who are happy, free, relaxing, respected and well looked after (all Chinese teachers);
- be more open (Teachers CN3, CN7 and CN9) and have low expectations (Teacher CN5);
- expect learners to be more independent but at the same time also welcome them to ask the teacher to teach them again if they do not get the knowledge taught (Teachers CN3 and CN4).

It seems that teachers from two countries imagined each other's teaching more about the role of the teacher, the way children might be grouped in class, and the level of mathematics they were possibly expected to learn. At the end of the interview, every teacher was asked about her/his possibility to change if there were other ways of maths teaching that might work. Nineteen teachers all said they were open to change, but there were apparent national trends. While teachers from England were more flexible and used to change, two-thirds of teachers from China were more concerned about issues that might prevent significant and/or fast change, such as the differences of cultures and educational systems. Of the three Chinese teachers who had not mentioned any obstacles to change, two were first-year teachers showing their willingness to actively learn anything new that could enrich their practices immediately.

To this point, we have attempted to seek deeper understandings about individual teachers' beliefs through hearing their responses to all the interview questions. Next, we will hear teachers' collective views on two maths lessons, one from England and the other from China, which they generated with their colleagues in the focus groups regarding the effectiveness of mathematics teaching.

Seeing maths teaching through groups of teachers' eyes

In each focus group, teachers saw two lessons, EN2 and CN7, and commented on the effectiveness of the lessons collectively. As follows, for each of the two lessons, first, the subject content and the lesson process will be briefly described; second, native and foreign colleagues' views will be presented and interpreted. Before dipping into detail, it should be pointed out that the length of each subsection was solely determined by the amount of information that each strand of data had to offer. For example, the focused English lesson did not consist of lots of teaching events, nor much change of classroom activities, so that part was relatively shorter than the part on the focused Chinese lesson which had considerably more events and thus demanded more words to be described.

The focused maths lesson from England: content and process

Given the fact that children had learned times tables in the previous year, Lesson EN2 was aimed to consolidate children's times tables with an extra attention from the teacher to a few low-ability pupils' 10 and 20 number bonds.

EN2 component-1: reasons and targets to learn

This lesson was the second in a sequence of lessons on the same content – Times Tables. At the beginning, Teacher EN2 spent 7 minutes on whole-class interaction in which the teacher first asked children to talk about the purpose of learning times tables and then showed them the Steps of Success Criteria on the IWB (Figure 7.2).

EN2 component-2: independent work and self-assessment

Then pupils were sent to independent work on coloured practice sheets. These sheets were readily organised in six clearly labelled transparent folders (Figure 7.3) on the wall, and children came and chose their own sheets from time to time. Such independent work lasted 40 minutes in which children self-timed their calculations, self-assessed their answers with calculators and allowed themselves to progress to the next-higher-level worksheet if they thought it had taken no more than a minute for them to complete the prior-level sheet correctly. While most pupils were working independently, the teacher and the TA were each working with a very small number of low-ability pupils.

EN2 component-3: self-assessing goals achieved and to be achieved

During the last four minutes of the lesson, the teacher asked pupils to read through the learning targets (i.e. Steps of Success Criteria), reflect upon their work, mark what they had achieved that day and plan what they would work on the next day. The lesson ended here, and the teacher started an English lesson without a break.

LO: I can calculate **Steps to success:** 1. I can calculate my number bonds to 10 and 20. 2. I can calculate my 2, 5 and 10 times tables. 3. I can calculate my 3 and 4 times tables. 4. I can calculate my 6 and 8 times tables. 5. I can calculate my 7 and 9 times tables. 6. I can use the inverse.	Our steps to success and our activities!
For each step you must: ✓ practise the skill (number bonds or times tables recall) and complete the worksheet. You can use any classroom resources you wish to help. . . ✓ check your answers using a calculator, make sure you correct any incorrect answers with a coloured pencil. Decide whether you are ready to test yourself or if you need to do the sheet again. ✓ complete a record card for your focus number bond and times tables whilst timing yourself. ✓ check your answers using a calculator. If you got them all correct and in under 1 minute, then move on! if you didn't get them all correct, then you must keep practising with the record card until you do!	

Figure 7.2 Success Criteria (lesson objectives) of Lesson EN2

Step to success number 1	*Step to success number 2*	*Step to success number 3*
Step to success number 4	*Step to success number 5*	*Step to success number 6*

Figure 7.3 Differentiating labels of the worksheets in Lesson EN2

Next, we will hear various opinions from teachers in the focus groups organised within each country regarding the quality and effectiveness of the focused English lesson. As introduced in previous chapters, there are two focus groups in each country, which makes a total of four focus groups: EN-FG1 (n=5), EN-FG2 (n=5), CN-FG1 (n=6) and CN-FG2 (n=6–3).

The focused maths lesson from England – native colleagues' views

Five English teachers in focus group 1 (EN-FG1) thought that Teacher EN2's lesson was ineffective:

> They (the pupils) were just practising skills to make them quicker . . . She (i.e. Teacher EN2) was there facilitating them doing sheets – that's how I saw it . . . There was a lot, at the beginning, of teacher talking to the whole class. But not an input teach[ing]. It wasn't giving them anything.

In the other English group which included Teacher EN2, teachers agreed during the group discussion time that the lesson offered more differentiation to children and were more pupil-led than the Chinese lesson. Nonetheless, they later on expressed their actual opinions about the English lesson to the researcher personally in their school:

> The teacher only spent ten minutes out of about one hour interacting with the whole class and there was no content going on . . . It was unbelievable that the English teacher took a whole lesson on times tables. The content was unbelievably easy. That English school should have set by maths.

Overall, all native colleagues but one (Teacher EN1) genuinely considered that the lesson EN2 was ineffective, time-consuming and short of teacher input. It would be interesting to hear what foreign colleagues reckon about the same lesson.

The focused maths lesson from England – foreign colleagues' views

Most Chinese teachers thought the English lesson offered children unlimited opportunities to explore their learning, to choose the level of worksheets that they felt they could do and to self-assess their learning, which were quite different from the Chinese maths class. They pointed out:

> The (English) lesson was more relaxing and individualised, and the children more independent . . . The teacher did not need much time to prepare lessons (i.e. do daily plans).

On the other hand, they also thought the English lesson had low expectation of children's calculation ability and had less content delivered over 50 minutes. Some teachers argued:

> If the final exam paper were the same to every pupil, then differentiated learning goals would mean nothing and be particularly unfair to those low-ability children, because they would get worse and worse, being taught like that.

Teacher CN4 thought that those children in the class were probably of the level of Grade 2 in China (i.e. Year 3 in England, meaning two years behind China). Discovering that the teacher spent quite a long time giving guidance to one pupil, some teachers talked to each other that none of the Chinese teachers would have done that because they had to get the content grasped by the majority rather than a single child over a limited amount of lesson time. In a quite straightforward way, Teacher CN8 said:

> This lesson was very time-consuming and thus not effective and it was impossible for Chinese teachers to adopt this approach while maintaining the teaching pace.

Nonetheless, the majority of teachers agreed that each country had its own standards, so depending on the English standards this lesson might be effective.

The following two sections show the content and process of the focused Chinese lesson and the collective views of each country's teachers on the focused Chinese lesson.

The focused maths lesson from China: content and process

The topic of the lesson was about numbers from *wans* (万, i.e. Chinese equivalent place name for 10 thousands) to thousand *wans* (千万, i.e. Chinese equivalent place name for 10 millions, see notes 2 and 3) and their place value. The aim of this lesson was to introduce the names of these four places and how to accurately read and write numbers that fell into this range. As aforementioned, the Chinese numeral system is slightly different from that of the West in that the former separates numbers by four digits whereas the latter by three. Thus, one place up the thousands place is given a Chinese name, *wan* (万), which is the equivalent of 10 thousands in English. For example, the number 10000 is read as "one *wan*".

CN7 component-1: leading in

The teacher started with the historical story of the way ancient people counted tens, hundreds and thousands by tying knots of different sizes. This story led to the idea of base 10. She then pointed out today's lesson was about "recognising numbers (认数)" while writing down this title on the board. She left a space between认 and 数 where she later on inserted three characters "识整万" (literally, knowing whole-*wan*) to form the complete title – 认识整万数 (Recognising & Knowing Whole *Wan* Numbers), immediately after the concept of whole-*wan* numbers had emerged in the whole-class discussion.

CN7 component-2: reviewing four places (units through thousands)

At the video time point 00:02:16, the teacher asked pupils about the name of each number place, and as they called out the column headings, she quickly wrote units (个), tens (十), hundreds (百) and thousands (千) from right to left on the board. Then the teacher showed a slide on which two numbers (1023 and 3002) were represented on abacuses which had four columns (i.e. units to thousands) of beads; through Q & A, the class reflected upon the reading and writing methods of the two numbers.

CN7 component-3: introducing new digits

At 00:04:14, the teacher suggested that the class manipulate their abacuses together with her to represent numbers thousand by thousand starting from 5000. Every time after adding a bead onto the thousands, there would be a round of Q & A about how to read and write the number and what the beads meant, for example, children would say that six beads on the thousands meant six thousands. After 9000, 1 0000 emerged, at which point the new place – *wans* (万， as aforementioned, the Chinese equivalent of 10 thousands) – was introduced, and the teacher wrote the place name to the left of the thousands on the board. The activity continued until the number reached a hundred *wan* (100 0000).

CN7 component-4: generating writing and reading methods

At 00:17:36, the teacher showed six numbers on a new slide (Figure 7.4) and initiated a round of Q & A about the thinking method for reading the numbers correctly. The most agreed answer was to count four zeros from the units up through higher places, prepare to read the group of four zeros as one Chinese character *wan* (万i.e. 10 thousand) and read aloud the remaining number on the left hand side first and the character "wan" second.

下面这些数你会读吗？ *(Can you read the following numbers?)*		
10000	40000	70000
90000	50000	30000

Figure 7.4 Task on a slide from Lesson CN7 (Component-4)

At 00:18:25, the teacher turned to the next slide which showed a picture and a description which read "Nanjing Olympic Stadium has a total of 6 0000 seats." She asked a pupil to read it for the class. Then, another slide came, asking pupils to write two numbers according to their readings in Chinese. Once done, the teacher asked pupils to check the answers with their desk-mates and talk to each other the writing methods they used. To conclude this task, the teacher initiated another round of Q & A for individuals to share with the whole class their methods.

CN7 component-5: generating the idea of the Level-separating Line

At 00:22:34, the teacher showed a slide on which four numbers were given. Then she posed two questions:

> Could you read these numbers? Are there any methods that could help us read the numbers after a quick glance?

Immediately, pupils started to volunteer to answer the questions. At the moment when a pupil pointed out the idea of grouping numbers by four digits, the teacher spotted an opportunity to introduce the Chinese numeral system where every group of four digits formed a level (Chinese equivalent of 'period'). She then wrote the names of two levels: the units level (个级， from units to thousands) and the *wans* level (万级， from *wans* (i.e. 10 thousands) to thousand *wans* (i.e. 10 millions)) (Figure 7.5). The idea of the Level-separating Line (分级线) was introduced following a relevant answer from a pupil. The teacher then drew the red dotted line on the board between the units level and *wans* level.

Immediately after this, she asked the class, "How shall we divide this number 50000 with the Level-separating Line?" Children answered the questions randomly. When some pupils said it should start from the units, the teachers asked the whole class, "Counting from the units, how many places up?" The class answered, "Four places!" The teachers asked, "Then what?" The class answered, "Draw a Level-separating Line."

On the board, the teacher drew a dotted line between 5 and the 0 next to 5 in 50000 and asked, "Once we draw the Level-separating Line, is it easy to read the number?" Pupils responded loudly, "Easy!" "How to read it?" asked the teacher. "Five *wan*!" While children were reading "five", the teacher simultaneously pointed at the number 5; while they were reading "*wan*", she quickly pointed across the remaining four zeros from the thousands down to the units. She tried to simultaneously complement her body language with children's reading.

After this, the method of drawing the line and then reading 6 0000 was carried out similarly but faster. At this point, the teacher suggested pupils to choose three numbers on their textbooks, try out the method independently and then discuss their methods with their desk-mates.

<table>
<tr><td rowspan="2">PowerPoint slide on the screen</td><td colspan="8">下面这些数你会读吗？(Can you read the following numbers?)</td><td rowspan="2">Step 1</td></tr>
<tr><td colspan="4">120000
200000</td><td colspan="4">150000
500000</td></tr>
<tr><td rowspan="2">Teacher notes on the board</td><td colspan="4">万级</td><td colspan="4">分级线
个级</td><td rowspan="7">Step 2</td></tr>
<tr><td>千万</td><td>百万</td><td>十万</td><td>万</td><td>千</td><td>百</td><td>十</td><td>个</td></tr>
<tr><td rowspan="5">Five pupils' work on the board</td><td></td><td></td><td></td><td>5</td><td>0</td><td>0</td><td>0</td><td>0</td></tr>
<tr><td></td><td></td><td></td><td>6</td><td>0</td><td>0</td><td>0</td><td>0</td></tr>
<tr><td></td><td></td><td></td><td>8</td><td>0</td><td>0</td><td>0</td><td>0</td></tr>
<tr><td></td><td></td><td>1</td><td>0</td><td>0</td><td>0</td><td>0</td><td>0</td></tr>
<tr><td>1</td><td>0</td><td>2</td><td>0</td><td>0</td><td>0</td><td>0</td><td>0</td></tr>
<tr><td rowspan="2">PowerPoint slide on the screen</td><td colspan="8">下面这些数你会读吗？(Can you read the following numbers?)</td><td rowspan="2">Step 3</td></tr>
<tr><td colspan="4">12 ¦ 0000
20 ¦ 0000</td><td colspan="4">15 ¦ 0000
50 ¦ 0000</td></tr>
</table>

Figure 7.5 Introducing the Level-separating Line in Lesson CN7

CN7 component-6: representing numbers with the level-separating line in mind

At 00:29:56, the teacher showed another slide asking the class to represent and count 10 *wan* (i.e. 100 thousand) by 10 *wan* from 'nine hundred and sixty *wan*' (960 0000) to 'one thousand and twenty *wan*' (1020 0000) on their abacuses, and then write the numbers on their workbooks. At 00:33:53, the teacher asked whether the Level-separating Line did help in writing numbers and, if yes, how. Individuals were asked to talk about their methods.

CN7 component-7: reviewing and consolidating the lesson content

At 00:30:04, the teacher pointed at the title of the lesson again and asked the class to look carefully for the similarities between the numbers shown on the slide. These numbers are 85 and 850000, 850 and 8500000, 8500 and 85000000, 805 and 8050000, 8005 and 80050000 and 8050 and 80500000.

After a round of Q & A, at 00:39:22, pupils were asked to turn to page 87 in the textbook and work on problem #3. After a few seconds, the teacher said she found some pupils were all doing the same thing. She then called a girl to tell the class what it was, and the girl said she was drawing the Level-separating Line. The teacher gave the girl a swift praise and recommended the method to the class.

Noticing that the class was all done, the teacher initiated a round of Q & A for individuals to stand up to read the numbers comparatively: e.g. eighty-five (85) vs eighty-five *wan* (85 0000), eight hundred and fifty (850) vs eight hundred and fifty *wan* (850 0000), and so on. Another round of Q & A concluded the reading and writing methods for whole-*wan* numbers[6] at the *wans* level. At 00:43:03, the teacher initiated the last round of Q & A regarding the question "what have you learnt today?" The lesson ended one and a half minutes later.

Next, we will hear teachers' views on this focused Chinese lesson which they expressed in the focus groups with colleagues.

The focused maths lesson from China – native colleagues' views

Chinese teachers across two groups came up with the agreements that this lesson had both a clear structure and a logical and coherent connection between sections of the lesson, that the correct rate of children's work would be very high for this lesson and that the lesson represented what the local teachers had been doing in the class. Teachers tended to dig deep into specific steps of the lesson. For instance, Teacher CN2 thought the lesson was taught with great detail and clarity, but the lesson content was not difficult so the teacher should not have allocated a long time for abacus manipulations and using Level-separating Lines. Teacher CN6, however, noted that the reason the teacher spent longer time on the learning of the Level-separating Line might be that the teacher was trying to give the children more time to explore and find the usefulness of drawing the Level-separating Line and consolidate their knowledge of place value. He thought it seemed helpful in developing children's ability.

Weaknesses of the lesson were also discussed at times. For example, in CN-FG1, while acknowledging the lesson was effective in achieving the knowledge targets, Teacher CN1 also argued that the lesson did not offer enough opportunities for children to explore the knowledge. In CN-FG2, Teacher CN8 suggested that the teacher might consider replacing the physical abacuses with practice sheets on which children could draw circles to represent beads. With this method, children would be able to gradually develop their abstract-thinking ability by relying less upon physical objects to represent numbers, given the fact that they were progressing towards senior years. Teacher CN7 tended to turn a critical eye on her own lesson. In CN-FG2, she volunteered to be the first to talk and pointed out several of her mistakes in the lesson: (1) mistakenly writing the word 计 (counting) as its homophone 记 (recording/writing); (2) writing the place of *thousand wans* (千万, 10 millions) on the board at a time which otherwise could have been a little earlier. She later on added that she should have introduced to the children the differences between the Western (3 digits) and Chinese (4 digits) ways of separating number places into periods.

The focused maths lesson from China – foreign colleagues' views

Regarding the negative aspects of the Chinese lesson, FG-EN1 teachers offered quite brief comments regarding the lack of differentiation and the teacher standing at the front, and they then went on discussing about its positive points for a long time. They were impressed by the clarity of the lesson, children's motivation, children's respect to teachers and their behaviours:

> The clarity of her explanation in the context was phenomenal. It was so clear. She had the visual of the abacuses that every individual child was doing. She had the visual on the board of the numbers moving. She had children coming up. She had the context. It's very simple, but it's really effective.
>
> . . .
>
> Such motivation! Such drive! In the English lesson, it was the teacher enthusing the children by setting tasks and motivating them by using the timer to beat their time. There is more responsibility on the part of children [in the Chinese lesson]. In England, it's all on the teacher's.
>
> . . .
>
> These children they come to the classroom, [and] they are so motivated to learn. Our children, if we stand in front of the classroom, it would be so difficult to motivate them. You know, we came in one day and we dressed up. We did a big show for them. Some of the children were like "So what? So what?" They just . . . They take it for granted. They don't appreciate the effort gone in. And even that the extent of effort on the teacher's part to make enthusiastic sometimes isn't enough.
>
> . . .
>
> There's interaction, paired work and assessment. They were very focused. They talked to their partners. Once they got the function, they turned straightaway and faced the teacher, and they're ready for the next bit. Not like in mine, when they finished talking, they'd talk about what was on telly last night.

All five teachers in EN-FG2 thought the Chinese lesson was very much driven by the teacher and totally teacher-led, with "no real independence at all" and no differentiation. Teacher EN3 said the teacher should do "prior-learning" to see whether the pupils knew it already, because he had a sense that they "kind of already knew it". Every time a teacher gave an opinion – most of the time a negative one, other teachers would immediately agree to it, saying, "Yeah." The only "positive" point they seemed to all hold was about the discipline, but this was expressed in a tone of joking. When Teacher EN4 said she "like" the way the pupils all did exactly what they were told. Teacher EN3 added:

> The behaviour? That was impeccable, wasn't it? To answer a question, was literally I put up my hand and I stand up, which is really interesting to see.

Three other teachers immediately responded with a smile and "yeah". It is observable that English teachers in either group were inclined to agree with their colleagues and did not attempt to offer different views from others within a group. Two clear distinctions between two English focus groups were

(a) EN-FG1 teachers were more willing to talk, whilst EN-FG2 teachers talked less on both focused lessons probably because of the presence of Teacher EN2 and (b) EN-FG1 teachers talked more about the positive aspects of the focused Chinese lesson whereas EN-FG2 teachers commented more on its negative aspects.

Chapter conclusion

In this chapter, we have heard a rich collection of voices regarding the effectiveness of mathematics teaching. These include the researchers' unstructured analysis of mathematics lessons, teachers' teaching beliefs and self-comments on their own lessons, their international awareness and attitudes to change and teachers' collective views on the two lessons, EN2 and CN7. Now that we have all data analysed, it is time to connect the dots to see how the initial results and findings are mutually related and how the integrated findings can illuminate future research, policy and practice.

Notes

1 *wan*: phonetic transcription (pinyin) of the Chinese character, 万, numerically equivalent to 10,000. In this book, we also use the plural form "*wans*" when representing corresponding Chinese place names (万位，十万位，百万位，千万位) and a specific four digit period in Chinese numerical system, 万级 (the *wans* level).

2 In Chinese numerical system, the *wans* level or period (万级) consists of four digits/places: 万位 *wans* (10 thousands), 十万位 ten *wans* (100 thousands), 百万位 hundred *wans* (millions) and 千万位 thousand *wans* (10 millions). See the English-Chinese corresponding places below:

English places	*10 millions*	*millions*	*100 thousands*	*10 thousands*
Chinese places	thousand *wans* 千万	hundred *wans* 百万	ten *wans* 十万	*wans* 万

3 The Chinese numerical system separates every four consecutive places, instead of three, into a period. For example, 1,000,000 is read out by separating the number like 100 0000.

4 A Level-separating Line functions similarly as a comma (e.g. in 10,000) in the Western numerical system. It is used informally as an aid for children to read large numbers easily.

5 The *yis* level (亿级) consists a group of four higher places sitting to the left (i.e. higher value) of the *wans* level (note 2). See the English-Chinese corresponding place names below:

English places	*100 billions*	*10 billions*	*billions*	*100 millions*
Chinese places	thousand *yis* 千亿	hundred *yis* 百亿	ten *yis* 十亿	*yis* 亿

Whole-yi numbers are numbers that have zero in places from the units through 10 millions. For example, 16 0000 0000 reads as 16 *yi*.

6 This refers to numbers whose last four digits (from units up to thousands) are exactly zeros, e.g. 4 0000.

8 Connecting the dots

Chapter overview

- Correlations between maths teaching and learning
- Multiple perspectives on effective maths teaching – one aim
- Connecting numbers with voices

The EMT project has collected data from 10 English teachers along with their 236 pupils and nine Chinese teachers together with their 343 pupils across mathematics classrooms in two cities that were each socio-economically equivalently positioned in their home countries, England and China. It applies six methods – structured lesson observations (M1), questionnaires (M2), standardised mathematics tests (M3), unstructured lesson observations (M4), video-stimulated interviews (M5) and video-stimulated focus groups (M6) – to address the three major research questions and their corresponding sub-questions:

- *RQ1: correlations between maths teaching and learning*
    - RQ1a: evaluating teacher behaviours
    - RQ1b: evaluating learning outcomes
    - RQ1c: correlating teaching with learning
- *RQ2: multiple perspectives on effective maths teaching*
    - RQ2a: the researcher's perspective
    - RQ2b: the teacher's perspective
    - RQ2c: native and foreign colleagues' perspectives
- *RQ3: interconnections between hard evaluations and soft voices*

The remaining part of this chapter will draw evidence from the results and findings of the study to address the research questions sequentially.

Correlations between maths teaching and learning

This section has three subsections, each focusing on a sub-question of RQ1, respectively.

The evaluation of teacher behaviours

The study evaluated the quantity and quality of teacher behaviours, as observed in nineteen maths lessons in Southampton and Nanjing, with two internationally validated observation systems – Opportunity to Learn (OTL) and International System for Teacher Observation and Feedback (ISTOF).

OTL results

OTL focuses on six percentages in relation to a lesson: the percentages of five types of classroom activities and the average percentage of pupils on task at regular intervals throughout the lesson. The five types of classroom activities are: *whole class interaction, whole class lecture, individual/group work, classroom management* and *partial class interaction.*

Pooling both countries' data together, the OTL evaluation found that English teachers allocated about a quarter (23.8%) of lesson time for whole class interactions and a half (46.6%) of the time for pupils working on their own without sufficient teaching input and that Chinese teachers spent about three quarters (72.2%) of lesson time interacting with the whole class, leaving about a quarter (27.8%) of the time for individual/group work. Whilst English teachers spent a varying amount of lesson time lecturing to the whole class (3.8%), managing the class (3.5%) or interacting with individuals or part of the class (22.3%), no Chinese teachers allocated any time for any of the three types of activities. Notably, the average time in which English teachers interacted with individuals or part of the class occupied approximately one-fifth of the lesson time, and the time for whole-class interaction was almost the same amount. In addition, on average, the proportion of English pupils on task was 92.8%, whereas almost all Chinese pupils (99.7%) were on task. Since the class size in China was larger, the number of pupils off task per class was much fewer in China than in England. Only one English classroom had 100% pupil time on task, with the percentages for the rest of English classrooms ranging from 85% to 99%. Conversely, just one pupil in each of two Chinese classrooms was off task over a short period, and all pupils in the other seven classrooms were on task throughout the lesson.

ISTOF results

The revised version of ISTOF focuses on six dimensions of 40 effective teacher behaviours: assessment and evaluation (ISTOF 1), clarity of instruction (ISTOF 2), instructional skills (ISTOF 3), promoting active learning and metacognitive skills (ISTOF 4), classroom climate (ISTOF 5) and classroom management (ISTOF 6). All teachers' behaviours are each evaluated on a five-point scale, ranging from "strongly disagree = 1" to "strongly agree = 5" and clustered around 19 effectiveness indicators.

In ISTOF 1, on average, English teachers score 12.8 (SD = 2.3), and Chinese teachers 19.4 (SD = 1); in ISTOF 2, English teachers get 16.4 (SD = 4.4), and Chinese teachers 27.9 (*SD* = 1.7); in ISTOF 3, the English mean is 11.4 (*SD* = 5.1), and the Chinese 23.2 (*SD* = 1.7); in ISTOF 4, English teachers reach 21.7 (*SD* = 6.9), and Chinese teachers 47.3 (*SD* = 2.9); in ISTOF 5, English teachers average 22.8 (*SD* = 6), and Chinese teachers 38.8 (*SD* = 1.1); in ISTOF 6, English teachers get 18.1 (*SD* = 6.4), and Chinese teachers 34.2 (*SD* = 1.7). Chinese teachers outperform their English colleagues in each of the six ISTOF domains. English teachers also have larger standard deviations than their Chinese colleagues, which suggest a larger within-country difference in England.

The evaluation of learning outcomes

In order to assess 9- to 10-year-old children's maths performance internationally, a 40-item test was derived from the released item bank of TIMSS 2003 Grade-4 maths which was designed for children at the same age. The test was constructed by strictly following the proportions of content and cognitive domains defined by TIMSS. The initial intention was to assess both learning outcomes and learning gains with two tests scheduled at the beginning and end of 10 school weeks in the second semester of the 2012–2013 school year. The results suggested a ceiling effect upon Chinese pupils in Test 2 which prevented the actual measure of Chinese pupils' improvement over 10 weeks. It is possibly because the items were too easy for Chinese children. The two test results were therefore correlated with teaching measures separately to evaluate teaching impact on learning outcomes at two points in time rather than gains over time.

The 40 TIMSS items included in the two EMT tests fall into five content domains – *Number* (n = 16), *Patterns and relationships* (n =6), *Measurement* (n = 8), *Geometry* (n = 6) and *Data* (n = 4) – and four cognitive domains – *Knowing facts and procedures* (n = 8), *Using concepts* (n = 7), *Solving routine problems* (n = 17) and *Reasoning* (n = 8). Correct rates (%) were calculated at the item, content domain, cognitive domain and paper levels. A pupil's paper-level correct rate stands for the overall test score for her/him.

At the item level

Each country's correct rate on every item is not only mutually compared but also compared with the TIMSS 2003 international average. In ***Test 1***, England has lower correct rates (*difference* ∈ [1%, 72%]) on 36 items than China and higher (*difference* ∈ [1%, 38%]) on 21 items than the TIMSS 2003 average; China has lower rates (*difference* ∈ [17%, 49%]) on 3 items than England, a same rate on item #17 as England's and lower rates (*difference* ∈ [10%, 11%]) on two items than the TIMSS 2003 average, exceeding the TIMSS average on 38 items (*difference* ∈ [7%, 67%]). In ***Test 2***, England's correct rates are lower (*difference* ∈ [4%, 62%]) on 35 items than China and higher (*difference* ∈ [1%, 45%]) on 31 items than the TIMSS 2003 average; China gets lower rates (*difference* ∈ [1%, 43%]) on five items than England and exceeds the TIMSS average on all 40 items (*difference* $\in$ [2%, 75%]). In either test, China has a higher correct rate than England on 35 or more items.

At the domain level

The two countries' correct rates in various domains are compared alongside the EMT international means. In ***Test 1***, England outperforms China in one of five content domains – *Data* – by 12% and has lower rates in the remaining four content domains (*difference* $\in$ [23%, 36%]) and all four cognitive domains (*difference* $\in$ [17%, 37%]) than China and in all content (*difference* $\in$ [12%, 18%]) and cognitive (*difference* $\in$ [9%, 19%]) domains than the EMT pooled average. In ***Test 2***, England still makes higher correct rates (*difference* = 12%) than China and the EMT average (*difference* = 6%) in *Data* but lower in other content (*difference* $\in$ [9%, 19%]) and cognitive domains (*difference* $\in$ [13%, 30%]) than China and lower than the EMT average (content domain *difference* $\in$ [10%, 14%], cognitive domain *difference* $\in$ [6%, 15%]).

At the paper level

For every country, the mean of all papers' correct rates and the spread of children's performance are calculated. In Tests 1 and 2, English pupils have achieved a mean of 56% (SD = 21) and 66% (SD = 19), respectively, and Chinese pupils averaged 83% (SD = 10) and 87% (SD = 10). In both tests, Chinese pupils outperformed their English peers by over 20% with approximately half of English standard deviation. Chinese pupils might have achieved an even higher mean in each test, if not for the context-specific issues in items #9 and #10 (see Chapter 6). The international differences on both occasions were statistically significant with a strong effect (Test 1: $p < 0.001$, $d = 1.74 > 1$; Test 2: $p < 0.001$, $d = 1.44 > 1$).

The overall results are in line with the identified gaps in mathematics performance between England and China in previous international studies (Lapointe et al., 1992; Lapointe et al., 1989; OECD, 2010, 2013).

Correlating teaching with learning

To evaluate the effects of teaching on children's learning outcomes in mathematics, every teacher's OTL percentages and ISTOF domain scores are correlated with the class mean in each of the two maths tests cross-nationally.

Correlating OTL results with maths performance

Pooling data from two countries together, the correlational analysis indicated that two OTL components had a significantly positive correlation with pupil performance in either mathematics test, and these components were *whole-class interaction* (r = 0.97 or 0.91, $p < 0.01$) and *pupil time on task* (r = 0.95 or 0.91,

$p < 0.01$). An OTL component – *whole-class lecture* ($r = -0.91$ or -0.92, $p < 0.01$) – posed a significantly negative effect upon mathematics scores in both tests, and two others – *individual/group work* ($r = -0.81$, $p < 0.05$) and *classroom management* ($r = -0.77$, $p < 0.05$) – were significantly related to lower scores in test one.

Correlating ISTOF results with maths performance

All six ISTOF components were found to correlate with higher maths scores in both tests. Pearson's *r* correlation coefficients of the six ISTOF components varied between 0.89 and 0.99 across two tests, the correlations between ISTOF measures and pupil maths scores were all significant at the level of 0.01. The greatest coefficients were of *promoting active learning and developing metacognitive skills* ($r = .99$ or .93) and *classroom climate* ($r = 0.97$ or 0.94).

Multiple perspectives on effective maths teaching – one aim

This section reflects upon various views regarding the effectiveness of maths teaching from the researcher, the teachers themselves and their local and international colleagues.

In the researchers' mind

The researchers' unstructured observations focus on the flow and the characteristics of each lesson. In either aspect, there emerges a strong sense of national patterns.

Lesson flow

English lessons all started from lesson targets – what teachers called Success Criteria, went through short and speedy first input and then long repetitive re-inputs, and ended in children's self-assessment against "Success Criteria". Chinese lessons all started from real-life situations, went through intensive whole-class interaction around two to three exemplary tasks each followed by one or two similar tasks for independent work and ended in real-life situations. In the English class, children's independent work took a considerably longer period than in the Chinese class where the teaching input took a relatively longer time and was carried out systematically and thoroughly so children could easily tackle similar problems quickly afterwards.

Lesson characteristics

In the *English maths class*, teachers (1) delivered differentiated teaching content, (2) shared all learning targets with the whole class at the very beginning of a lesson, (3) tended to transmit the knowledge directly to children, (4) re-taught individuals what had been just taught as many times as being asked for, (5) were facing children with weak foundation and poor readiness for the new content, (6) tended to care less about the quality and accuracy of their handwriting on the board while demonstrating procedures and/or solutions, (7) allowed pupils to move frequently and freely in the lesson particularly during independent work and (8) had at least one teacher assistant available for special educational needs (apart from classes EN3, EN4, and EN10).

In the *Chinese maths class*, teachers (1) made way for pupils to think actively and independently in finding solutions, (2) asked lots of questions to externalise children's thinking and reasoning of mathematical problems and of mathematical thinking and reasoning itself (metacognition), (3) were facing children with solid foundation and absolute readiness for the new content, (4) utilised pupils' classwork as teaching resources, (5) took seriously the timing of their writing on the board, i.e. when to write what, (6) spent most of the lesson time on teacher-guided and pupil-centred whole-class discussion, (7) facilitated less physical movement of pupils during the lesson and (8) had a big class size with little support for special educational needs.

In the maths teacher's mind

Individual interviews with teachers collected teachers' perspectives in three main aspects: teacher beliefs in EMT, self-evaluation of the observed lesson and international awareness and attitude to change.

Teacher beliefs in EMT

CRITERIA OF AN EFFECTIVE MATHS LESSON

English teachers thought an effective maths lesson should be relevant to real life, differentiated and child-centred, whilst Chinese teachers thought it should be aimed for and well accepted by pupils, be well planned, have real-life and long-term meaning and have realised lesson targets.

STRATEGIES OF ORGANISING AND MANAGING LESSONS

English teachers all pointed out the strategy of assessment-aided differentiation, whereas Chinese teachers offered more approaches: paired discussion, the use of warm reminder, jokes or riddles, specific strategies for specific needs of each lesson, promoting peer instruction among pupils and being well prepared and flexible.

STRATEGIES FOR DIFFERENTIATION

English teachers talked about four ways of differentiation: self-differentiation, pre-assessment, support from the TA and setting different tasks for different-ability children; Chinese teachers aimed for the majority, applied questioning techniques to offer opportunities for all-ability children, gave post-lesson support to low-ability and special awards to high-ability pupils and might suggest low-ability pupils to preview the lesson content.

Teacher self-evaluation of the observed lesson

LESSON PLAN

Apart from Teachers EN1 and EN2 who both planned to teach times tables, all other English teachers did not have the same planned lesson content. Chinese teachers were more likely to share the same teaching content in and across schools, probably due to the existence of Lesson Plan Groups within each school and the unified textbooks within each province.

THE IMPLEMENTATION OF THE LESSON PLAN

Whilst all teachers thought their lesson plans were implemented well, many of them also reflected upon some flaws. English teachers were more likely to find their intended content did not match the levels of some or all children; Chinese teachers were more likely to find imperfection in small steps of the process which they thought as crucial for nurturing logical thinking. The central issue for English teachers seems to be *what* to teach and to teach *whom what*, whilst that for Chinese teachers is likely to be *how* to teach. The former is rooted in the level of curriculum content and the corresponding differentiation, and the latter behaviours of teaching specific content. The former offers *certain* children opportunities to learn *certain* knowledge at *certain* levels, and the latter focuses on the *coherent* development of children's thinking in mathematics and offers the *same* learning opportunities to *all*.

THE EFFECTIVENESS OF THE OBSERVED LESSON

All teachers thought their lessons were effective, and all but two English teachers (EN1 and EN2) also spotted something imperfect. Again, as shown in their reflections upon their lesson plans, English teachers were more concerned with differentiation, whereas Chinese teachers were more worried about the function of specific teaching steps.

International awareness and attitude to change

THE IMAGINED FOREIGN LESSON

English teachers imagined that Chinese pupils would sit in rows either individually or in pairs, face the front, not be allowed to move and be well-motivated, on task and hard-working, that Chinese lessons would be very much teacher-led and structured and that there would be textbooks and more advanced content.

Chinese teachers imagined that English pupils would be enjoying more freedom, sitting around tables in groups and receive plenty of individual care, that English lessons would be more relaxing and have looser standards and that the English class would be quite open and have a smaller class size.

TEACHER FLEXIBILITY TO CHANGE

All teachers expressed their willingness to change, but, at the same time, two-thirds of Chinese teachers also pointed out various things that might prevent them from thinking of change immediately and tremendously, such as the adaptability of successful approaches from other cultures in their own culture.

Overall, Chinese teachers tended to reflect on what had not been done well and know why it was that and how it might be improved immediately in the next lesson. English teachers were also concerned with certain aspects that were not ideal, but it seemed that those aspects, mostly extrinsic – such as differences among pupils' ability, had existed for long in their classrooms, hence seemingly impossible to change immediately.

In the local and foreign colleagues' mind

By presenting two video-recorded lessons – one from England and the other from China – to teachers in four focus groups across the two geographical locations, local and foreign colleagues' views were collected regarding the effectiveness of the two lessons.

Collective views on the maths lesson from England

Apart from Teacher EN2 and her colleague Teacher EN1, all other English teachers thought that this lesson was time-consuming and that teaching was almost absent. On the other hand, Chinese teachers had more diverse views on it. Out of nine Chinese teachers, three thought the lesson was ineffective and would lead to poor results; three regarded it as child-centred and individualised; two feared that this way of teaching would be unfair for low-ability children and would widen the performance gap among children over time; two saw the lesson as relaxing; one thought it could be effective if evaluated with the English standards; one thought this approach was rather time-consuming.

Collective views on the maths lesson from China

Half of English teachers thought the Chinese lesson was too much teacher-led and lacked differentiation; all of them thought the classroom discipline was impressive; the other half thought the Chinese children were self-motivated, eager to learn, and very focused and that the Chinese lesson was effective even though it had much fewer resources than the English – "it's very simple, but it's really effective." Chinese teachers talked a lot about details, interconnections between details, and the impact of the lesson upon children's development in and beyond the learning of mathematics. All agreed that the lesson would be effective in terms of correct rates, almost all pointed out both pros and cons of the lesson and reflected upon similar dilemmas they faced in practice, some gave suggestions about ways of optimising specific steps of the lesson, and some even scrutinised possible reasons for certain phenomena in the lesson.

Overall, English teachers' views were clustered within groups, whereas Chinese teachers' views were diverse and complementary to each other and their thinking was professionally deeper and more critical.

Connecting numbers with voices

This section addresses the third research question by interconnecting key results and findings from quantitative and qualitative parts of the study.

Differentiation and ability gaps (EN)

The fact that English teachers spent an average of 22.3% (*SD* = 25.4%) of lesson time on partial class interaction fits in English teachers beliefs in the importance of teaching differentiation. In the English classroom, the constant streaming of children during the teaching process offered fewer opportunities for certain children to learn certain content, which over time might widen the performance gap. The much wider performance gap is evident in English children's test results, which is also in line with the ability picture that English teachers sketched in the teacher questionnaire (see Chapter 5).

Pro direct transmission, re-inputs and partial interaction (EN)

English teachers were pro direct transmission, which was evident in the teacher questionnaire and unstructured lesson observations. Because they took the direct transmission approach, their cycle of delivering a set of knowledge was quick – too quick to get children's thinking thoroughly developed. The majority of children just didn't get it, and the speedy first input did not work, so the teacher had to re-do the teaching time and time again to individuals or part of the class in the same way. As the model of teaching did not fundamentally change, the following re-inputs did not seem to work well either.

Child-led or not? (EN)

The universally practised Success Criteria across the English classrooms echoes English teachers' strong belief in the child-led explorative approach. However, this belief contradicts with their beliefs in both the direct transmission of knowledge and the differentiation of teaching through individual interactions, in that the actual teaching does not allow children to lead the thinking process.

In the English classroom, it is the teacher that holds the standard answer/solution and demonstrates it to the class. Therefore, when all these beautiful ideas come together in practice, the intended learning simply does not happen.

Teaching for all, pro-constructivist and whole class interaction (CN)

Chinese teachers' focus on all level children as expressed in interviews is consistent with the fact that they have spent an average of 72.2% (*SD* = 9.3%) of lesson time *interacting* with the whole class. Also, as found in the questionnaire data, they are pro constructivists, which can be triangulated by their teaching processes. The use of whole-class interaction and discussion offers more opportunities for them to externalise and develop children's mathematical thinking thoroughly on one task before moving on to the next. The intensive questioning and answering in the whole class enriches children's thinking and reasoning experiences, with them frequently explaining their ideas or listening to others' ideas, which allows children to mutually construct knowledge throughout the lesson. It reflects teachers' constructivist beliefs (as shown in Teacher Questionnaire) – that underpin their teaching processes – [TQ #29*l*] *thinking and reasoning processes are more important than specific curriculum content* and [TQ #29f] *pupils learn best by finding solutions to problems on their own*. The fact that the belief and practice of TQ #29*l* works is not because the curriculum content is unimportant, but because mathematics in itself is so highly coherent and logical that it is only through thinking and reasoning of it that children can get the content. During the thinking and reasoning processes in the whole class, the guidance of the teacher is the key to effectiveness and the *catalyser* to the development of children's thinking within limited lesson time. The Chinese way of teaching puts the mission of finding knowledge tightly in the children's hands instead of the teacher's. The Chinese belief and practice of letting children find solutions on their own (TQ #29f) is neither through children's independent work nor through the teacher's direct delivery of intended content. Rather, it is through the teacher-guided child-centred approach[1] in the course of whole class interactions.

The effectiveness of first input and teachers' SMK and PCK

English teachers spent less time on the first input of knowledge, which could be partly explained by their lack of sufficient subject and pedagogical content knowledge. The questionnaire data also confirmed that English teachers were not specialist maths teachers and that they, however, expressed a lower level of needs in developing their subject matter and pedagogical content knowledge than did Chinese teachers who were all specialist teachers. Chinese teachers all spent longer time on the first occasion of knowledge input, which was not only observable in all lessons but also interpretable in teacher interviews. For instance, Teacher CN1 said, "I must make sure every pupil clearly understand the method through the teaching and learning of ***one*** example . . . " In the centre of a Chinese lesson were intensive whole-class interactions and discussions around no more than four examples which were in a hierarchical order and did not simply repeat each other. This is consistent with the finding that more effective teachers have limited content focus per lesson in the Junior School project (Mortimore et al., 1988). Chinese teachers ensured that children's *first* opportunity to learn must be successful, which echoes with similar findings of high success rates in early American TER studies (Berliner et al., 1978; Fisher et al., 1978) and with similar principles held by Chinese maths teachers in Ma's work (1999, p. 46) who "pay particular attention to the first time an idea is introduced to students in its simplest form".

On the contrary, English teachers' first input did not result in pupils' mastery of the delivered knowledge. Although they thereafter gave children multiple opportunities to learn the same content, the ineffectiveness seemed always there – as if a shadow to the unchanging teaching processes. Because of English teachers' lack of subject matter and pedagogical content knowledge, the first input was not well structured, the big aim was not broken into digestible smaller chunks, the lesson components did not interconnect mutually coherently, the time applied on each occasion of input was too short for children to think thoroughly, and the questions from the teacher lacked necessary depth. Such a connection between teacher knowledge and teaching quality is evident in the literature (e.g., Ma, 1999; Tikunoff et al., 1975).

Lesson objectives, direct transmission and constructivist

The way two countries' teachers present lesson objectives also reflects their beliefs. All English teachers presented directly and clearly, at the very beginning of each lesson, the content and targets that would be covered and were expected to achieve. This falls in the format of direct transmission. Conversely, most Chinese teachers did not clarify the lesson targets at the beginning. Rather, they started from Q & A about real-life issues or phenomena and pointed out the title of the lesson when relevant answers emerged, the detailed knowledge points would appear one by one on either a slide or the chalkboard as they taught, and the lesson objectives gradually entered the scene one at a time as the lesson went on.

Self-reflection and metacognition

Both countries' teachers facilitated pupils' self-reflection upon learning, but there were considerable differences. English teachers put more emphasis on reflecting upon the lesson goals and results – what I'll achieve today (at the beginning), what I have achieved today, and what I aim to achieve tomorrow (at the end). Nonetheless, Chinese teachers stressed more on the process and reasons behind it – how you have done it, how someone has made that mistake and why, what is the reason for learning this, and so forth. English teachers often asked pupils to make reflections at the beginning and the end of the lesson, while Chinese teachers kept prompting pupils to do so throughout the lesson. It is thus not surprising that Chinese teachers scored higher in ISTOF in promoting metacognitive skills, because thinking of thinking processes – rather than thinking of learning outcomes (the *right* facts) – is the key to developing metacognition.

Teacher questioning: who solves the problem?

The most significant distinction between the English and the Chinese approaches perhaps lies in the way of teacher questioning. Typical scenarios would be that the former threw a question to one or several individual(s) and soon chose to explain the answer or procedure her/himself if the 'correct' answer did not emerge, whereas the latter not only asked more questions but also constantly prompted the children to find the answer(s)/solution(s) themselves through intensive questioning and answering. In the English classroom, there is a standard procedure or solution that the teacher feels must be transmitted to the children; in the Chinese classroom, as aforementioned, there are multiple ways of solving a problem which are expected to emerge in the lesson *from* pupils. Chinese teachers were very *unlikely* to tell the pupils the "right" solution to *any* problem – big or small, even though the solution was the core target of the lesson. Chinese pupils generated and explained each solution during the process of teacher-initiated Q & A. The purpose of such intensive whole-class discussion and exploration was to get pupils to arrive at the lesson destination(s) by themselves. English teachers generally held the solution and passed it on to pupils directly while interacting with them. If all pupils could apply the solutions to similar problems, then the English lesson was deemed successful.

Who's in charge of what?

English teachers were in charge of what should be the correct solution/procedure, though they let children be in control of the level of work they would like to work on. Whilst Chinese teachers were more likely to be in charge of the lesson rhythm/process, they let the children lead the way and think, find and tell every possible solution/procedure to every problem. In English maths lessons, there seemed to be one possible solution to a given problem; in Chinese maths lessons, there were many possible solutions to a single problem. In English maths lessons, the teacher held the unique key to a specific unknown and passed it on to children so that they could also unlock it in the same way; in Chinese maths lessons, the children held the key to trying multiple possible keys to an unknown with varying forms in varying situations. In the English maths lessons, the teacher played the main role of mathematical thinking and reasoning; in the Chinese maths lessons, it was the pupils that played this role.

Who repeats what?

English teachers tended to re-teach multiple times in a lesson the same content after the first input, whereas Chinese teachers were more likely to ask pupils to find the solution(s) and then re-explain it/them at different points in the lesson.

Where to conclude?

English lessons all had a final main conclusion in the end, while Chinese lessons all had a mini-conclusion near the end of each discussion throughout the lesson process and might or might not have a final conclusion at the end of the lesson.

Classroom management and corresponding CPD needs

English teachers on average spent 3.5% of lesson time managing the class but had lower percentages of pupils on task; Chinese teachers did not spend any time on classroom management but had higher percentages of pupils on task. Combining findings in observations and interviews, one could easily see that Chinese teachers were more capable of managing the class without apparent managing behaviours than English teachers. Nonetheless, in the questionnaire, Chinese teachers demanded more needs in developing their classroom management skills than did English teachers. It may be because they appreciate the importance of management skills and have been trying hard to improve such skills that they actually turn out to be better at it, which is similar to the case of subject matter and pedagogical content knowledge.

The differences in the amount and content of CPD might indirectly influence how teachers teach, and this is probably all CPD programmes intend to do. However, the maths class is the arena where formal teaching and learning of mathematics happens, and children learn maths from teachers' teaching directly rather than their CPD. To improve maths teaching, designers of primary maths CPD programmes may consider reinforcing the effective teaching factors identified in the EMT study. School leaders should create an environment for more school-based CPD to happen and nurture a positive climate in which teachers are happy to get involved and continue to grow.

Cultural explanations

Researchers from the TIMSS Video Study come up with the idea that teaching is a *cultural* activity (Stigler & Hiebert, 1999). The EMT study proves that they are both right and wrong. They are right because the process of mathematics teaching varies considerably from England to China. They are wrong because they neglect the fact that culture is a collective habit of thinking and doing things and thus fail to conclude such differences at both macro and micro levels. At the macro level, it does seem to suggest that teachers in different cultures teach mathematics in different ways. However, at the micro level, the end products of a culture in the classroom are various elements of teaching and learning. So, is it that the differences of cultures make a difference? Or is it that the varying proportions of different teaching and learning elements make a difference? The EMT evidence points us to a positive answer to the latter.

Moreover, culture is not static; it is dynamic and constantly evolving. Members of a certain culture are influenced by the culture they perceive and believe in and at the same time are also constantly reformatting/reconstructing the culture with or without clearly defined purposes. Certainly, schools and classrooms also have their own mini-cultures, and the change of institutional cultures has increasingly been attempted by teachers and school leaders (Fullan, 2007). Cross-cultural studies are undoubtedly eye-opening, as they offer alternative options from other cultures and generate new ideas and methods in the process of observations and reflections.

Homework differences

The results of the pupil questionnaire show that children from two cities have a different amount of homework, but research suggests that the amount of homework has approximately zero impact on pupils' achievement in primary schools (Cooper, 1989). Through meta-analyses, Cooper (1989) identified differentiated effects of the *amount* of homework on pupils' achievement at different schooling stages: (1) $r \approx 0$ for Grades 3 through 5 (Years 4 through 6 in England), (2) $r = 0.07$ for Grades 5 through 9 (Years 6 through 10 in England) and (3) $r = 0.25$ for upper secondary school. It was recommended that the amount of weekly homework should be (i) one to three assignments (each lasting no more than 15 minutes) for lower primary pupils, (ii) two to four assignments (each lasting 15 to 45 minutes) for upper primary pupils, (iii) three to five assignments (each lasting 45 to 75 minutes) for lower secondary pupils, and (iv) four to five assignments (each lasting 75 to 120 minutes) for upper secondary pupils.

Though studies do suggest a positive effect of homework on achievement (Cooper, Robinson, Patall, 2006; Good & Grouws, 1979), the positive effect will not happen unless the homework is of an optimal amount and good quality and has reasonable purposes. Otherwise, the effect can be negative (Hallam, 2004). Comparing with time on task and other factors, the effect size of homework also tends to be smaller (*ibid.* as cited in Muijs & Reynolds, 2011, p. 142).

In summary, the amount of homework will have little impact on children's learning if they have not grasped thoroughly the knowledge underpinning their homework. In the EMT study, children in the English classroom kept looking for re-inputs from the teacher, which had already indicated the ineffectiveness of teaching. The amount of homework would mean nothing to lots of (in some classes almost all) English pupils after class. In the Chinese classroom, children spent the majority of lesson time gradually understanding the knowledge through deeper thinking and actively answering interconnected questions; their seatwork was quickly and evenly checked by the teacher; work samples with right or wrong solutions were picked up, presented and discussed in the whole class. After class, the Chinese pupils demonstrated a thorough understanding of the knowledge taught and they were ready for independent homework regardless of the amount. Without the mastery of knowledge in the class, the amount of homework would add little positive value to pupils' maths performance.

Discussion

Overall, English teachers tended to emphasise the importance of differentiation and the idea of child-centred teaching; however, they seemed to have over contradicted whole-class interactive teaching with the child-centred idea. As a result, they seemed to uniformly believe that putting children at the centre of the class could only be realised through individualised interactions. Chinese teachers on the other hand were more concerned with detailed steps of teaching and saw the steps of teaching as a coherent, irreversible and logical whole, just like the hierarchical nature of mathematics itself. They also put the children at the centre of the teaching and learning, but their emphasis was on the children taking the lead of the problem-solving process in the scope of the whole class. The teacher-pupil and pupil-pupil interactions in the whole class demanded everyone's full attention, as the questioning and answering was both brisk and engaging. Their approach is more of scaffolding and constructivism than direct transmission, whereas the English approach is more of direct transmission – which is surprising but is evident across the 10 English classrooms. English teachers see the differences among pupils as an obstacle and therefore feel the necessity of teaching them separately according to their levels; Chinese teachers see the differences among pupils as natural resources of teaching and learning, that contribute to the progress of everyone in the class, and thus are quite comfortable about putting children together and interacting with them actively.

It occurs to us that Chinese maths teaching is more progressive than the methods applied in Britain and many other Western countries. Both caring about every individual pupil (equity and democracy), the English maths teachers spend more time teaching individuals rather the whole class and allocating about a half of the lesson time for children to work on their own, but the Chinese maths teachers stick to active interaction with the whole class in about three-quarters of the lesson time, breaking the remaining lesson time into a few small segments for independent work and inserting them purposefully between whole-class interaction/discussion periods. The English approach means that there are always some children being excluded from the interaction with their teacher, whereas the Chinese methods ensure that every single pupil is included in the interactive teaching process.

In the English maths class, it might seem that the teacher gives the children entire freedom, but the focus on interaction with individuals and independent work before children can be *truly* independent means that some children are from time to time excluded from the democracy of equal learning opportunities. In the Chinese maths class, it might appear that all individuals are most of the time taught together, but it is essentially through whole-class interaction with the teacher that their rights to learn the same content as everybody else are equally respected and realised.

Both believing in the importance of hands-on activities through which children learn by doing, English teachers, despite apportioning a huge chunk of lesson time for pupil "independent" work, have to re-teach many individuals who at times get stuck and choose to come back to their teachers throughout the practising time; Chinese teachers, apparently seeing the thinking and reasoning process as the heart and soul of hands-on (maybe more precisely: minds-on) work for *mathematics*, embed such hands-on activities naturally into the whole class interaction time, and structure the interaction with logically interconnected mini-questions. Through answering lots of questions during the whole-class interaction time, Chinese children's mathematical thinking becomes highly observable, which in turn constantly feeds back to the teacher and guides the dynamics of the teaching and learning process.

English maths teaching is more about the teacher demonstrating the correct way of doing mathematics and pupils learning to do it in the same way. Chinese maths teaching is more about the teacher developing the mathematical knowledge gradually in pupils' minds through intensive interaction with the whole class.

Chinese maths teachers seem to have found a better path towards a truly progressive education, with the characteristics that Dewey proposed a century ago realised, such as hands-on, child-centred, teacher-guided, teacher as a facilitator (rather than a demonstrator), relevant to children's life. Maybe the Western world has been misinterpreting – rather than understanding and developing – Dewey's thoughts on education from the very beginning.

Chapter conclusion

We have finally reached the destination and found the answers to the questions we set out to look for at this point where the beginning met the end. In the EMT study, we have scrutinised the teaching and learning reality and captured its details from different angles. Those pieces of reality have finally met each other again to represent the complete whole reality of maths teaching and learning, which this chapter has explicitly witnessed. Carrying the details and the completeness of the reality, we now move to the final chapter to conclude the study with lessons and suggestions for future research on and improvement of mathematics teaching in and across nations.

Note

1 This approach is what Chinese teachers commonly call "以教师为主导，学生为主体".

9 Improving maths teaching cross-nationally

Chapter overview

- What we have achieved thus far
- We could have achieved more
- Improve maths teaching cross-nationally, now!
- Final remarks

Looking back and forward, in this final chapter, we share with you what we have achieved thus far on this cross-national research journey, what are the things that we could have done but haven't in the EMT study, what we will and other researchers should be doing in the future and most importantly what strategies we would like to offer for everyone in primary maths education – be you teachers, school leaders, policymakers or researchers – to improve maths teaching cross-nationally, now!

What we have achieved thus far

The study drew on both quantitative and qualitative strands of data collected in mathematics classrooms for 9- and 10-year-olds in two socio-economically equivalent cities from two cultures and two countries. It correlated teacher behaviours with mathematics learning outcomes, captured multiple perspectives from different roles regarding the effectiveness of mathematics teaching within and across national borders and scrutinised the interconnections between hard measurement (science) and soft voices (human). It advances the fields of teaching effectiveness research and mathematics pedagogical research in but not limited to the following eight aspects:

- collecting cross-national evidence on effective maths teaching in primary schools;
- replicating effective maths teaching 'elements' found in previous national studies;
- discovering TER evidence from the West in an Eastern context;
- first TER study attempting the value-added approach across nations;

- applying both teaching effectiveness and improvement methods;
- first TER study seeking to form an international dialogue on EMT;
- reuniting two communities of maths teaching research;
- the integrated paradigm, new methods, the holistic reality;
- methodological contribution: multifunction of videos in TER.

A cross-national voyage towards effective maths teaching

The study filled in the gap in the literature where international comparisons lacked sufficient attempts to interconnect teaching and learning in one study, where the evidence of TER studies was mainly drawn from national studies and barely crossed nation borders, and where hard evaluations of correlations between classroom variables lacked the crucial support from "soft" (as often regarded) but rich explanations that only qualitative data may offer. It is the first study after ISERP (Reynolds et al., 2002a) that has ever collected TER data cross-nationally and brings to light in time cross-national empirical evidence that the world is longing to know regarding what works in the teaching of mathematics internationally. It is also the first TER study that has ever attempted to evaluate the value that different teaching approaches might add to pupils' mathematics performance *cross-nationally*.

Effective maths teaching formula – we know it all along

The results and findings of the study further confirm what has been repetitively found as effective at the classroom level in previous studies conducted within different countries. It reassures that certain teaching factors do travel internationally as foreseen by the ISERP research team (Reynolds et al., 2002a). It consolidates the foundation of TER and the development of educational science in the subject area of mathematics cross-nationally. It tests the adaptability of two established observational instruments – OTL and ISTOF – in two geographically and culturally considerably different countries. The pooled analyses of two countries' data coincide with both the underlying hypotheses of the two instruments and the existing evidence in the literature.

Trace the OTL correlates back to the literature

A high opportunity to learn has already been found in more effective teachers' classrooms about four decades ago in that it guarantees both quantity and quality of children's engagement into the academic content that they are learning (Arehart, 1979; Fisher et al., 1980; Stallings, Cory, Fairweather, & Needles, 1977). The study by Good and Grouws (1979) indicated that whole class instruction could only be effective if the teacher taught in an *interactive* way. This is evident in this study's OTL correlates in that whole class *interactive* teaching predicted higher maths performance whereas whole class *lecture* led to lower maths performance. The British study, MEPP by Muijs and Reynolds (2003), also found the positive effect of whole class interactive teaching upon both learning outcomes and gains, which is consistent with partial findings in previous TER studies in the UK and US (Croll, 1996; Emmer et al., 1979; Galton et al., 1980; Pollard et al., 1994). The literature also suggests that teachers who spend more time interacting with the whole class tend to have higher rates of time on task (Croll, 1996; Muijs & Reynolds, 2003; Pollard et al., 1994). Conversely, teacher interaction with individual pupils poses a negative effect upon *time on task* (Croll & Moses, 1985) which is, however, crucial in nurturing pupils' thorough engagement into subject learning rather than social activities and promoting optimal learning outcomes/gains (Muijs et al., 2014; Muijs & Reynolds, 2003).

Trace the ISTOF correlates back to the literature

More effective teachers are good at asking questions to assess the status of pupils' understanding and guide them to continue thinking and reasoning through follow-up questioning, which is also evident in the literature (Evertson et al., 1980; Muijs et al., 2014). A lesson with a clear and logical structure follows the way the human brain processes information, hence contributing to the quality of teaching (Clarke et al., 1979; Land, 1979). Effective instruction happens in the classroom where the teacher engages and interacts with *all* pupils through appropriate questioning and answering and thus develops their understanding of intended academic knowledge (Fisher et al., 1980; Stallings & Kaskowitz, 1974). The effective way of promoting active learning and metacognitive skills is also mainly through teacher questioning. Effective teachers tend to ask more process questions which lead children to more active reasoning and critical thinking (Evertson et al., 1980; Muijs et al., 2014; Muijs & Reynolds, 2000). More effective teachers are good at building a learning-oriented climate through consistently effective management of time and pupils' learning momentum and steadily effective interactions with all pupils (Brophy, 1973; Muijs et al., 2014; Muijs & Reynolds, 2000). More effective teachers are good at managing time, classroom activities and pupil behaviours (Berliner & Tikunoff, 1977; Good & Grouws, 1977); the EMT study further suggests that they are capable of preventing off-task misbehaviours from happening through highly interactive teaching with the whole class, and thus no apparent managing actions are needed.

The discovery of TER evidence from the West in an Eastern context

The majority of effective teaching elements have been long discovered in the West using the teaching effectiveness research approach. They are for the first time discovered in China – a context that is generally believed to be quite different from the West. This reassures that certain teaching factors do travel across nations and cultures.

First TER study ever attempting the value-added approach across nations

The study employs pre- and post-tests with a genuine intention to assess the academic value that teachers might have added to pupils' learning in maths over time (10 school weeks). It is also the first study that has ever conducted cross-study comparisons of English children's maths performance using exactly the same items – 40 TIMSS 2003 maths items, which offers a historical view of the performance of 9- to 10-year-olds in both content and cognitive dimensions over the course of nine years.

Advancing the TER field by applying effectiveness and improvement methods

Inheriting the traditional quantitative traits of TER, the study develops the field through incorporating both the teaching effectiveness (QUAN) and teaching improvement (QUAL) approaches. The study sees both approaches as parts of a whole rather than two mutually contradicting sides.

First TER study seeking to form an international dialogue on teaching

Across the nation border, the study has listened to multiple voices of individuals and groups of teachers about the same topic – what works in teaching maths in primary schools. This is the first TER study that has ever attempted to collect and appreciate practitioners' views across nations and cultures. About maths teaching in general and the effectiveness of two focused lessons in particular, teachers' voices have travelled across space and time and joined each other in this international dialogue in the EMT study.

Reunion of two communities of maths teaching research

This is not only one of the few TER studies that have attempted to understand the ways in which teachers handle the subject matter but also one of the few maths pedagogic studies that have systematically looked into what teachers *do* in the maths class. The EMT study interweaves all strands of data and evidence around maths teaching and appreciates both the content and the process of teaching. When it comes to maths teaching and learning, there are no either-ors – there have to be both. The study is another starting point for maths education researchers and effective maths teaching researchers to reconcile and work together for a better maths education for every child.

New paradigm, new methods, the holistic reality

The EMT project starts from the mixed-method paradigm and arrives at a new paradigm – *the integrated paradigm* – that makes use of merits of both positivism and interpretivism yet distinguishes itself sharply from the mixed-method approach in that it does not simply mix different methods together. Rather, it interweaves evidence from various angles – numbers and voices, connects perspectives from various roles – the researcher, the teacher, the native colleagues and the foreign colleagues – and interrelates research questions and answers in different layers at different points in the research process. The essence of this paradigm is that the world is an inseparable whole where the holistic *external* reality and the individually perceived *internal* realities inter-exist and interrelate.

The integrated paradigm offers new ways of designing, conducting and analysing research to researchers in education and other social sciences who are still cornered by the dilemma of choosing *either* positivism *or* interpretivism *or* a simple mixture of the two in one. This paradigm is helpful for both the research project itself and the subsequent dissemination of generalisable findings. For the project, the integrated paradigm is able to answer what (often QUAN) the researched issues are as discovered and why (often QUAL). After the project, people are able to develop sensible ways to improve the situation, and resolve issues, taking both workable solutions and practitioner perceptions into consideration.

Methodological contribution: multifunction of videos in TER

The study demonstrates an optimal utilisation of video in classroom observations, post-lesson interviews and focus groups. The use of video might not occur to be foreign for researchers in TER and other communities in educational research. However, the use of video for teacher reflections and collective comments appear to be rare in the current TER community, partially because of its traditional interest in quantitative data. In the EMT study, video-recorded lessons have made the international dialogue on the effectiveness of maths teaching easily come true, with teachers sitting in their comfort of local schools. Moreover, video data have provided the study an opportunity to integrate different types and strands of evidence. The study has thus set an encouraging example for future TER studies.

We could have achieved more

In this section, we reflect upon the limitations of the project and point out possible directions for research in this field in the foreseeable future.

A few regrets on this journey

The main limitations of the study lie in its sample size and statistical methods. It would have come up with richer findings if more schools from more regions in both countries had participated. The correlational method did not take other factors into consideration, such as the possible indirect impact of teacher professional development and/or teachers' subject knowledge on teaching and learning, the effect of homework marking and feedback on learning. Contextual variables, such as the degree of parental involvement, have not been included. Thus, the strength of a correlational study is its focus on key causes predicting specific results, which also becomes its weakness in that other essential factors and indirect predictors may have as well been ruled out.

The test instrument derived from TIMSS 2003 turned out to be too easy for the Chinese pupils. Subsequently, a ceiling effect has prevented the study from measuring learning gains. With a better set of test items, the study may have come up with more robust evidence for maths teaching improvement across nations in the long run.

Moreover, due to time limit and other conditions that might have limited a PhD study, only one lesson in each classroom was observed. Although the teachers were asked to deliver their lessons in their usual ways, a sequence of lessons from each class would have in themselves generated the typicality of the teacher's maths teaching.

In addition, the content of the observed maths lessons was not exactly the same, due to the difficulty of organising uniform teaching content across schools and nations. It would have generated deeper findings if all lessons were about the same subject area(s). Similarly, with the lesson content controlled as the same across classrooms and the test items designed for the same content as well, the study would have come up with effective teaching factors for specific maths topics for children at the specific age.

Last but not least, though every effort has been made, the study is still limited by the fact that only one researcher – the first author (then a PhD student) under the supervision from the second author – was independently carrying out this cross-national investigation. A team of researchers would have been able to generate more meanings from the multiple types of data. Though every effort has been made to keep the data collection and interpretation as neutral as possible, an international team would be much more helpful and efficient in working against potential biases.

A lot to look forward to next

Future TER studies are demanded with larger samples of participants recruited, more countries involved and multilevel modelling methods applied. Teaching improvement research projects are also much needed to bring TER evidence into practice, re-test teaching effects on each sample, and bring educational researchers, teachers and school leaders together to make evidence-informed innovations in classrooms. Researchers may consider experimental designs that involve pre- and post-lesson, pre- and post-unit, pre- and post-term assessment to compare the effects of different teaching approaches with the control of teacher-level variables, such as subject matter knowledge and length of teaching experience, and pupil-level variables, such as prior performance and family background.

Classroom-level factors are largely not investigated across the East and the West. The reason that international studies across different cultures and countries are meaningful and worth conducting is different nations have different educational policies and realities which could be seen as naturally existing laboratories where different educational experiments are constantly carried out (IEA, 2012). Different experimental inputs lead to different outcomes (*ibid.*). Studies across a variety of countries offer more opportunities to find which experiment works and which doesn't and to consolidate the knowledge base of educational science than studies within and/or across culturally similar countries. Given the typical performance gap between East Asian countries and Western countries, systematic large-scale TER studies will make meaningful contributions to the world's understanding of how to teach maths more effectively and scientifically.

This study only looked at the classroom level, so inter-country comparisons were only based on classroom-level data. There is more to be explored in the future to see correlations between independent and dependent variables at the pupil, classroom, school and system levels and between independent variables within and across these levels.

Improve maths teaching cross-nationally, now!

The performance gap in mathematics between the West and East Asia has been revealed repetitively in various international surveys of student achievement over the past couple of decades. Building on data collected from two cities that were each positioned in a socio-economically equivalent place in their home countries, this study further confirmed the existence of such a gap between English and Chinese 9- and 10-year-olds. Policymakers from the West are increasingly concerned with the results of international assessments and have been seeking ways of improving educational outcomes nationwide. However, most innovative plans tend to stop at the national or school level, hence missing the key lever to change – teaching at the classroom level.

In England, practitioners and policymakers have been heavily influenced by the advocates for progressive and individualised teaching (Central Advisory Council for Education, 1967), a broader curriculum, the superficial likes and dislikes of international comparisons (Alexander, 2012) and the one-sided assertion that pedagogy is merely an individualised and cultural thing (Alexander, 2008). The curriculum needs to be broad, which, nevertheless, cannot be an excuse for the lack of depth and effectiveness of curriculum implementation. The intended content of every subject, including mathematics, should be thoroughly and firmly grasped by children after all efforts have been put in at the system, school and classroom levels.

The danger is not what Pasi Sahlberg superficially calls "GERM" (as applauded by Alexander, 2012). The real danger is the lack of standards on education and its evaluation and the unawareness or ignorance of ineffective educational processes. There will not be any improvement if educational practitioners, researchers and policymakers are ready to play ostrich whenever encountering disappointing outcomes.

It is fine to maintain a broad curriculum (reasonably broad, not simply broad*er*, time is limited). Nonetheless, England should, at the same time, ensure the high quality of teaching and learning of the core subjects – such as English and mathematics – which form the base of many other subjects and children's more fulfilling future life in a more advanced world. When children are under-performing in these core subjects, it is time for all educational stakeholders to reflect, research and react on the effectiveness and the improvement of teaching and learning in these subjects. It is as simple as that. To solve the issue, one needs to look at and focus on the issue itself.

Empirical studies in the literature and the EMT study all suggest that the English pedagogy is severely individualised, despite a small reduction over an unbelievably long time span. The proportion of time in which children work on their own has dropped from two-thirds of the lesson time in the ORACLE study (Galton et al., 1980) to 46.6% in the EMT study. The percentage of time that English teachers spend interacting with the whole class has doubled since the late 1980s, jumping from 12% in the Junior School project (Mortimore et al., 1988) up to 23.8% in the EMT study.

In addition to the historical and contemporary advocates for progressive and individualised teaching, another reason for the heavy use of individualised activities in the English classroom might be that English teachers' lack of profound subject knowledge limits their confidence in interacting with the whole class and restrains their ability in breaking the intended main topic into small, digestible and hierarchical (as mathematics itself) chunks and organising questioning and reasoning activities in a coherent manner, with the whole class involved and engaged.

In fact, whether to interact with individuals or the whole class is ultimately not an either-or question when it comes to teaching and learning in the classroom. Each individual child matters – there is no doubt. However, before we could afford to (even if we could, do we really have to?) allocate a subject expert teacher for every single child and also before we could all decide to ignore the very fact that human beings are a social species who grow and evolve in social dialogues and activities not only as individuals but also as parts of a whole, a considerable amount of lesson time should be allocated for the teacher *interacting* (**not** lecturing) with the whole class, if we are looking for the maximum progress of *every* individual child. In the dynamic of whole class interactive teaching and learning, individuals are open to *all* ideas and solutions that could possibly be generated by the whole class, hence progressing more rapidly and holistically.

Findings also suggest the difference of the subject matter and pedagogical content knowledge between two countries' maths teachers. English teachers are generalists and their subject matter and pedagogical content knowledge is weaker than their Chinese counterparts who are mathematics specialists. This is evident in the way they deal with the content in the classroom. There is a necessity for English primary schools to arrange specialist teachers across subjects so that teachers are more capable of optimising the quality and widening the scope of their interactions with the class. Specialist teachers will also have more time and space to prepare and reflect upon lessons in their specialised subjects, and children will benefit the most from what every teacher is most capable of.

It is part of the research rationale of the study to collect internationally applicable effective teaching behaviours in mathematics classrooms. The pooled analysis of lesson data from both countries means that those effective teaching approaches work both within and across countries. Those approaches are worth trying out (or keeping for the case of China) in either country's mathematics classrooms.

In England, effort needs to be put into increasing the proportion of time on whole-class interactive teaching, reducing that on individual work or teacher interactions with individuals and avoiding whole-class lectures, so that every individual can benefit the most from every lesson. For English teachers and policymakers, the big challenge/dilemma lies in two aspects: (1) whole-class interactive vs partial class interactive or individual work and (2) specialist vs generalist teachers in primary school.

To deal with the dilemma, England needs to make choices. Recommendations based on decades of maths TER literature and the EMT project findings are given here for English practitioners and policymakers to consider:

1 Avoid the utilisation of whole-class lectures;
2 Increase the proportion of lesson time on whole-class interactions;
3 Reduce the proportion of lesson time on individual/group work;
4 Reduce the proportion of lesson time on partial class interactions;
5 Enhance teachers' subject matter and pedagogical content knowledge;
6 If possible, transform primary teachers into specialist teachers;
7 Make change in initial teacher education accordingly.

In China, attention should be placed on preserving the current teaching approaches in an increasingly Westernised (Americanised) trend in the country's education and society, increasing support to children with special education needs and preventing radical change from happening in the classroom. The problem in China does not lie in classroom teaching and learning. It lies in the political, social and cultural domains of the country.

Final remarks

The replicated findings about what works in the mathematics classroom make a wakeup call to practitioners, policymakers, educational improvement researchers and all other educational stakeholders in both countries and beyond, regarding the possibility of making change in the teaching and learning of mathematics. The multiple perspectives among teachers indicates the connections between teachers' beliefs and their behaviours, which offer possible indirect ways of improving teaching and learning, for instance, encouraging teachers to enhance their subject matter and pedagogical content knowledge, come out of their comfort zones and try out new approaches.

There is also a need to increase collaboration between researchers and practitioners for, but not limited to, three reasons: (1) this will promote the speedy circulation of research evidence between researchers and teachers, (2) this may nurture critical thinking amongst teachers and (3) researchers will get quick access to most updated information of what is going on in the classroom 'right now', therefore being more capable of helping identify and tackle key issues in the classroom.

Crucial things that policymakers can do are to look at both initial teacher education and continuing professional development of in-service teachers, to implement teaching innovations based on existent empirical evidence and to encourage and support more classroom studies for them to consistently evaluate what works, how and why and also for them to steadily feed fresh evidence back to practitioners, policymakers and other educational stakeholders.

References

Alexander, R. (2008). Culture, dialogue and learning: Notes on an emerging pedagogy. In N. Mercer & S. Hodgkinson (Eds.), *Exploring talk in school* (pp. 93–114). London: Sage.

Alexander, R. (2012). *Visions of education, roads to reform: The global educational race and the Cambridge Primary Review*. Retrieved 20 July 2015 from www.robinalexander.org.uk/wp-content/uploads/2012/04/Alexander-Chile-GMU-lecture-1-121030.pdf

Anderson, L., Evertson, C., & Brophy, J. (1979). An experimental study of effective teaching in first-grade reading groups. *Elementary School Journal, 79*(4), 193–223.

Arehart, J. (1979). Student opportunity to learn related to student achievement of objectives in a probability unit. *Journal of Educational Research, 72*(5), 253–269.

Askew, M., Rhodes, V., Brown, M., William, D., & Johnson, D. (1997). *Effective teachers of numeracy: Report of a study carried out for the Teacher Training Agency*. London: King's College, University of London.

Atkinson, R. C., & Shiffrin, R. M. (1968). Human memory: A proposed system and its control processes. In W. S. Kenneth & S. Janet Taylor (Eds.), *Psychology of learning and motivation* (Vol. 2, pp. 89–195). London: Academic Press.

Becker, I. L. (1978). *Instructional ingredients for the development of beginning reading competence*. Pittsburgh, PA: Learning Research and Development Center, University of Pittsburgh.

Becker, W. C. (1977). Teaching reading and language to the disadvantaged – What we have learned from field research. *Harvard Educational Review, 47*(4), 518–543.

Bennett, N. (1987). Changing perspectives on teaching-learning processes in the post-Plowden era. *Oxford Review of Education, 13*(1), 67–79. doi:10.1080/0305498870130107

Berliner, D., Fisher, C., Filby, N., & Marliave, R. (1978). *Executive summary of beginning teacher evaluation study*. San Francisco, CA: Far West Laboratory.

Berliner, D., & Tikunoff, W. (1976). The California beginning teacher evaluation study: Overview of the ethnographic study. *Journal of Teacher Education, 27*(1), 24–30.

Berliner, D., & Tikunoff, W. (1977). Ethnography in the classroom. In G. Borich & K. Fenton (Eds.), *The appraisal of teaching: Concepts and process* (pp. 280–290). Reading, MA: Addison-Wesley.

Bill & Melinda Gates Foundation. (2012). *Measures of effective teaching*. Retrieved 12 October 2012 from www.metproject.org/index.php

Boaler, J., Wiliam, D., & Brown, M. (2000). Students' experiences of ability grouping – disaffection, polarisation and the construction of failure. *British Educational Research Journal, 26*(5), 631–648. doi:10.1080/713651583

Boonen, T., Van Damme, J., & Onghena, P. (2013). Teacher effects on student achievement in first grade: Which aspects matter most? *School Effectiveness and School Improvement, 25*(1), 126–152. doi:10.1080/09243453.2013.778297

Borg, W. (1980). Time and school learning. In C. Denham & A. Lieberman (Eds.), *Time to learn* (pp. 33–72). Washington, DC: National Institute of Education.

Bosker, R., & van der Velden, R. K. W. (1989). *The effects of schools on the educational career of disadvantaged pupils*. Paper presented at the 2nd International Congress for School Effectiveness and School Improvement, Rotterdam, NL.

Boyer, C. (1968). *A history of mathematics*. New York: John Wiley & Sons.

Brophy, J. E. (1973). Stability of teacher effectiveness. *American Educational Research Journal, 10*(3), 245–252.

Brophy, J. E., & Evertson, C. (1976). *Learning from teaching: A developmental perspective*. Boston, MA: Allyn and Bacon.

Brophy, J. E., & Good, T. L. (1986). Teacher behaviour and student achievement. In M. C. Wittrock (Ed.), *Handbook of research on teaching* (3rd ed., pp. 328–375). New York: Palgrave Macmillan.

Cai, J. (2003). Investigating parental roles in students' learning of mathematics from a cross-national perspective. *Mathematics Education Research Journal, 15*(2), 87–106. doi:10.1007/bf03217372

Cai, J., Ding, M., & Wang, T. (2014). How do exemplary Chinese and U.S. mathematics teachers view instructional coherence? *Educational Studies in Mathematics, 85*(2), 265–280. doi:10.1007/s10649-013-9513-3

Cai, J., Moyer, J. C., & Wang, N. (1999). Parental roles in students' learning of mathematics: An exploratory study. *Research in Middle Level Education Quarterly, 22*(3), 1–18.

Cao, Z., Bishop, A., & Forgasz, H. (2007). Perceived parental influence on mathematics learning: A comparison among students in China and Australia. *Education Studies in Mathematics, 64*(1), 85–106.

Central Advisory Council for Education. (1967). *Children and their primary schools: A report*. London: H.M.S.O.

Centre for Cities. (2015). *Southampton factsheet*. Retrieved 10 April 2015 from www.centreforcities.org/city/southampton/

Chen, C., & Stevenson, H. (1995). Motivation and mathematics achievement: A comparative study of Asian-American, Caucasian-American, and East Asian high school students. *Child Development, 66*(4), 1215–1234.

Chen, X., & Li, Y. (2010). Instructional coherence in Chinese mathematics classroom – A case study of lessons on fraction division. *International Journal of Science and Mathematics Education, 8*(4), 711–735. doi:10.1007/s10763–009–9182-y

China Knowledge. (2014). *Nanjing city information*. Retrieved 19 May 2014 from www.chinaknowledge.com/CityInfo/City.aspx?Region=Coastal&City=Nanjing

Clark, R. E. (1982). Antagonism between achievement and enjoyment in ATI studies. *Educational Psychologist, 17*(2), 92–101. doi:10.1080/00461528209529247

Clark, R. E. (1989). When teaching kills learning: Research on mathematics. In H. N. Mandl, N. Bennett, E. de Corte, & H. F. Freidrich (Eds.), *Learning and instruction: European research in an international context* (Vol. 2, pp. 1–22). London: Pergamon.

Clarke, C., Gage, N., Marx, R., Peterson, P., Stayrook, N., & Winne, P. (1979). A factorial experiment on teacher structuring, soliciting, and reacting. *Journal of Educational Psychology, 71*(4), 534–552.

Clarke, D., Keitel, C., & Shimizu, Y. (Eds.). (2006). *Mathematics classrooms in twelve countries: The insider's perspective*. Rotterdam, The Netherland: Sense Publishers.

Clarke, D., Mesiti, C., O'Keefe, C., Xu, L. H., Jablonka, E., Mok, I. A. C., & Shimizu, Y. (2007). Addressing the challenge of legitimate international comparisons of classroom practice. *International Journal of Educational Research, 46*(5), 280–293. doi: http://dx.doi.org/10.1016/j.ijer.2007.10.009

Cohen, L., Manion, L., & Morrison, K. (2007). *Research methods in education*. (6th ed.). London: Routledge.

Coleman, J. S., Campbell, E., Hobson, C., MacPartland, J., Mood, A., Weinfeld, F., & York, R. (1966). *Equality of educational opportunity*. Washington, DC: US Government Printing Office.

Cooper, H. (1989). Synthesis of research on homework. *Educational Leadership*, *47*(3), 85–91.

Cooper, H., Robinson, J. C., & Patall, E. A. (2006). Does homework improve academic achievement? A synthesis of research, 1987–2003. *Review of Educational Research*, *76*(1), 1–62. doi:10.3102/00346543076001001

Crawford, J., Gage, N. L., Corno, L., Stayrook, N., Mitman, A., Schunk, D., . . . Newman, R. (1978). *An experiment on teacher effectiveness and parent assisted instruction in the third grade*. Stanford, CA: Center for Educational Research at Stanford.

Creemers, B. P. M. (1994). *The effective classroom*. London: Cassell.

Creemers, B. P. M., & Kyriakides, L. (2006). Critical analysis of the current approaches to modelling educational effectiveness: The importance of establishing a dynamic model. *School Effectiveness and School Improvement*, *17*(3), 347–366. doi:10.1080/09243450600697242

Creemers, B. P. M., & Kyriakides, L. (2008). *The dynamics of educational effectiveness: A contribution to policy, practice and theory in contemporary schools*. London: Routledge.

Creemers, B. P. M., Kyriakides, L., & Sammons, P. (2010). *Methodological advances in educational effectiveness research*. London: Routledge.

Croll, P. (1996). Teacher-pupil interaction in the classroom. In P. Croll & N. Hastings (Eds.), *Effective primary teaching* (pp. 14–28). London: David Fulton.

Croll, P., & Moses, D. (1985). *One in five: The assessment and incidence of special educational needs*. London: Routledge and Kegan Paul.

De Fraine, B., Van Damme, C., & Onghena, P. (2007). A longitudinal analysis of gender differences in academic self-concept and language achievement: A multilevel latent growth approach. *Contemporary Educational Psychology*, *32*(1), 132–150.

den Brok, P., Brekelmans, M., & Wubbels, T. (2004). Interpersonal teacher behaviour and student outcomes. *School Effectiveness and School Improvement*, *15*(3–4), 407–442.

de Grift van, W. J. C. M. (2013). Measuring teaching quality in several European countries. *School Effectiveness and School Improvement*, *25*(3), 295–311. doi:10.1080/09243453.2013.794845

Department for Education and Employment (DfEE). (1997). *Excellence in schools*. London: Her Majesty's Stationery Office.

Department for Education and Employment (DfEE). (1998). *The implementation of the national numeracy strategy: The final report of the Numeracy Task Force*. London: DfEE.

Dewey, J. (1916). *Democracy and education*. New York: Palgrave Macmillan.

DfE. (2013). *Schools in Southampton LA*. Retrieved 15 August 2013 from www.education.gov.uk/inyourarea/schools/lea_852_schools.shtml

Driessen, G., & Sleegers, P. (2000). Consistency of teaching approach and student achievement: An empirical test. *School Effectiveness and School Improvement*, *11*(1), 57–79. Retrieved from www.tandfonline.com/doi/abs/10.1076/0924-3453%28200003%2911%3A1%3B1-A%3BFT057

Dunkin, M., & Biddle, B. (1974). *The study of teaching*. New York: Holt, Rinehart & Winston.

Edmonds, R. (1979). Effective schools for the urban poor. *Educational Leadership*, *37*(1), 15–27.

Einsiedler, W., & Treinies, G. (1997). Effects of teaching methods, class effects, and patterns of cognitive teacher-pupil interactions in an experimental study in primary school classes. *School Effectiveness and School Improvement*, *8*(3), 327–353.

Emmer, E., Evertson, C., & Brophy, J. E. (1979). Stability of teacher effects in junior high classrooms. *American Educational Research Journal*, *16*(1), 71–75.

Emmer, E. T., Evertson, C., Sanford, J., & Clements, B. S. (1982). *Improving classroom management: An experimental study in junior high classrooms*. Austin, TX: Research and Development Centre for Teacher Education, University of Texas.

Evertson, C., Anderson, C., Anderson, L., & Brophy, J. E. (1980). Relationships between classroom behaviours and student outcomes in junior high mathematics and English classes. *American Educational Research Journal, 17*(1), 43–60.

Fisher, C., Berliner, D., Filby, N., Marliave, R., Cahen, L., & Dishaw, M. (1980). Teaching behaviours, academic learning time, and student achievement: An overview. In C. Denham & A. Liberman (Eds.), *Time to learn* (pp. 7–32). Washington, DC: National Institute of Education.

Fisher, C., Filby, N., Marliave, R., Cahen, L., Dishaw, M., Moore, J., & Berliner, D. (1978). *Teaching behaviours, academic learning time and student achievement: Final report of Phase III-B, beginning teacher evaluation study*. San Francisco, CA: Far West Laboratory.

Flanders, N. (1970). *Analysing teacher behaviour*. Reading, MA: Addison-Wesley.

Flanders, N., & Simon, A. (1969). Teacher effectiveness. In R. L. Ebel (Ed.), *Encyclopedia of educational research* (4th ed., pp. 1423–1431). New York: Palgrave Macmillan.

Fortune, J. (1967). *A study of the generality of presenting behaviours in teaching preschool children*. Memphis, TN: Memphis State University.

Fullan, M. (2007). *The new meaning of educational change*. (4th ed.). London: Routledge.

Gage, N. (1965). Desirable behaviors of teachers. *Urban Education, 1*(2), 85–95.

Gagné, E. D. (1985). *The cognitive psychology of school learning*. Boston, MA: Little Brown and Co.

Gagné, R. (1970). *The conditions of learning*. New York: Hole, Rinehart and Winston.

Galton, M., Hargreaves, L., Comber, C., Wall, D., & Pell, A. (1999). *Inside the primary classroom: 20 years on*. London: Routledge.

Galton, M., & Simon, B. (Eds.). (1980). *Progress and performance in the primary classroom*. London: Routledge & Kegan Paul.

Galton, M., Simon, B., & Croll, P. (1980). *Inside the primary classroom*. London: Routledge.

Givvin, K. B., Hiebert, J., Jacobs, J. K., Hollingsworth, H. H., & Gallimore, R. (2005). Are there national patterns of teaching? Evidence from the TIMSS 1999 Video Study. *Comparative Educational Review, 49*(3), 311–343.

Goldstein, H. (2003). *Multilevel models in educational and social research*. (3rd ed.). London: Edward Arnold.

Good, T. L., & Grouws, D. A. (1975). *Process-product relationship in fourth-grade mathematics classrooms*. Columbia, MO: University of Missouri, College of Education.

Good, T. L., & Grouws, D. A. (1977). Teaching effects: A process-product study in fourth grade mathematics classrooms. *Journal of Teacher Education, 28*(3), 49–54.

Good, T. L., & Grouws, D. A. (1979). The Missouri mathematics effectiveness project: An experimental study in fourth-grade classrooms. *Journal of Educational Psychology, 71*(3), 355–362.

Good, T. L., Grouws, D. A., & Beckerman, T. (1978). Curriculum pacing: Some empirical data in mathematics. *Journal of Curriculum Studies, 10*, 75–81.

Good, T. L., Grouws, D. A., & Ebmeier, H. (1983). *Active mathematics teaching*. New York: Longman.

Griffin, G. A., & Barnes, S. (1986). Using research findings to change school and classroom practice: Results of an experimental study. *American Educational Research Journal, 23*(4), 572–586.

Grisay, A. (1996). *Evolution des acquis cognigifs et socio-affectifs des eleves au cours des annees de college (cognitive and socio-affective development during the years in the middle school)*. Liege: Universite de Liege.

Gu, L. (1994). *Theory of teaching experiment: The methodology and teaching principle of Qingpu*. [In Chinese 青浦实验的方法与教学原理研究]. Beijing, China: Educational Science Press.

Hallam, S. (2004). *Homework: The evidence*. London: Institute of Education, University of London.

Hampshire County Council. (2013). *2011 Census headline facts and figures Southampton*. Retrieved 15 August 2013 from www3.hants.gov.uk/2011_census_southampton_summary_factsheet.pdf

Hart, K., Brown, M. L., Küchemann, D. E., Kerslake, D., Ruddock, G., & McCartney, M. (1981). *Children's understanding of mathematics: 11–16*. London: John Murray.

Hattie, J. A. (2003). *Teachers make a difference: What is the research evidence?* Background paper to invited address presented at the 2003 ACER Research Conference, Carlton Crest Hotel, Melbourne, Australia, October 19–21, 2003.

Hiebert, J., Gallimore, R., Garnier, H., Givvin, K. B., Hollingsworth, H., Jacobs, J., . . . Stigler, J. (2003). *Teaching mathematics in seven countries: Results from the TIMSS 1999 Video Study*. Washington, DC: National Center for Educational Statistics.

Hiebert, J., & Stigler, J. W. (2000). A proposal for improving classrooms teaching: Lessons from the TIMSS Video Study. *The Elementary School Journal*, *101*(1), 3–20.

Hiebert, J., & Stigler, J. W. (2004). A world of difference. *Journal of Staff Development*, *25*(4), 10–15.

Hill, H. C., Rowan, B., & Ball, D. L. (2005). Effects of teachers' mathematical knowledge for teaching on student achievement. *American Educational Research Journal*, *42*(2), 371–406.

Hill, P., & Rowe, K. (1995). Use of multi-level modelling in procedures for maximizing between school comparability of final year school-based assessments. *Multilevel Modelling Newsletter*, 7(1), 12–16.

Hill, P., & Rowe, K. (1996). Multilevel modeling in school effectiveness research. *School Effectiveness and School Improvement*, 7(1), 1–34.

Hodgen, J., & Askew, M. (2007). Emotion, identity and teacher learning: Becoming a primary mathematics teacher. *Oxford Review of Education*, *33*(4), 469–487. doi:10.1080/03054980701451090

Hodgen, J., Küchemann, D. E., & Brown, M. (2014). Learning experiences designed to develop algebraic thinking: Lessons from the ICCAMS project in England. In P. C. Toh, T. L. Toh, & B. Kaur (Eds.), *Learning experiences that promote mathematics learning* (pp. 171–185). Singapore: World Scientific.

Hodgkin, L. (2005). *A history of mathematics: From Mesopotamia to modernity*. Oxford: Oxford University Press.

Howe, C., Luthman, S., Ruthven, K., Mercer, N., Hofmann, R., Ilie, S., & Guardia, P. (2015). Rational number and proportional reasoning in early secondary school: Towards principled improvement in mathematics. *Research in Mathematics Education*, 17(1), 38–56. doi:10.1080/14794802.2015.1019914

Huang, R. (2002). *Mathematics teaching in Hong Kong and Shanghai: A classroom analysis from the perspective of variation*. Unpublished PhD thesis, University of Hong Kong.

Huang, R., Mok, I. A. C., & Leung, F. K. S. (2006). Practising in the Mathematics classrooms in China. In D. Clarke, C. Keitel, & Y. Shimizu (Eds.), *Mathematics classrooms in twelve countries: The insider's perspective* (pp. 263–274). Rotterdam, The Netherland: Sense Publishers.

Huggins, R., & Thompson, P. (2013). *UK competitiveness index 2013*. Retrieved 10 April 2015 from www.cforic.org/downloads.php

Hunter, M., & Russell, D. (1981). Planning for effective instruction: Lesson design In M. Hunter & D. Russell (Eds.), *Increasing your teaching effectiveness*. Palo Alto, CA: The Learning Institute.

Husén, T. (1967). *International study of achievement in mathematics: A comparison of twelve countries.* (Vol. 1–2). Stockholm: Almqvist & Wiksell.

IEA. (2007a). *Mathematics concepts and mathematics items.* Retrieved 21 February 2012 from http://nces.ed.gov/timss/pdf/TIMSS4_Math_ConceptsItems.pdf

IEA. (2007b). *TIMSS 2003 mathematics items: Released set (Fourth Grade).* Retrieved 21 February 2012 from http://timss.bc.edu/PDF/T03_RELEASED_M4.pdf

IEA. (2012). *Brief history of IEA: 50 years of educational research.* Retrieved 1 July 2012 from www.iea.nl/brief_history.html

Ilonca, H. (2001). *The relationship between the use of representations and instructional discourse in mathematics tasks.* Paper presented at the Annual Meeting of the American Educational Research Association (AERA), Seattle, US.

Increasing Competence and Confidence in Algebra and Multiplicative Structures (ICCAMS) Maths. (2015). *Survey results.* Retrieved from http://iccams-maths.org/survey/

Ingvarson, L., Beavis, A., Bishop, A., Peck, R., & Elsworth, G. (2004). *Investigation of effective mathematics teaching and learning in Australian secondary schools.* Canberra, Australian: Council for Educational Research.

Jacobs, J., Hiebert, J., Givvin, K. B., Hollingsworth, H., Garnier, H., & Wearne, D. (2006). Does eighth-grade mathematics teaching in the United States align with the NCTM "standards?" Results from the TIMSS 1996 and 1999 Video Studies. *Journal for Research in Mathematics Education, 37*(1), 5–32.

James, H., Stigler, J. W., Jacobs, J., Givvin, K. B., Garnier, H., Smith, M., . . . Ronald, G. (2005). Mathematics teaching in the United States today (and tomorrow): Results from the TIMSS 1999 Video Study. *Educational Evaluation and Policy Analysis, 27*(2), 111–132.

Jones, K., Mooney, C., & Harries, T. (2002). Trainee primary teachers' knowledge of geometry for teaching. *Proceedings of the British Society for Research into Learning Mathematics, 22*(2), 95–100.

Keet, A. (2012). *England.* Retrieved 22 August 2013 from www.england.org.za/

Kirschner, P. A., Sweller, J., & Clark, R. E. (2006). Why minimal guidance during instruction does not work: An analysis of the failure of constructivist, discovery, problem-based, experiential, and inquiry-based teaching. *Educational Psychologist, 41*(2), 75–86. doi:10.1207/s15326985ep4102_1

Konstantopoulos, S., & Sun, M. (2013). Are teacher effects larger in small classes? *School Effectiveness and School Improvement, 25*(3), 312–328. doi:10.1080/09243453.2013.808233

Konu, A. I., Lintonen, T. P., & Autio, V. J. (2002). Evaluation of well-being in schools: A multilevel analysis of general subjective well-being. *School Effectiveness and School Improvement, 13*(2), 187–200.

Kounin, J. (1970). *Discipline and group management in classrooms.* New York: Holt, Rineheart & Winston.

Krantz, S. G. (2010). *An episodic history of mathematics: Mathematical culture through problem solving.* Washington, DC: The Mathematical Association of America.

Kulik, J. A., & Kulik, C-L. C. (1992). Meta-analytic findings on grouping programs. *Gifted Child Quarterly, 36*(2), 73–77. doi:10.1177/001698629203600204

Kyriakides, L., Creemers, B. P. M., & Antoniou, P. (2009). Teacher behaviour and student outcomes: Suggestions for research on teacher training and professional development. *Teaching and Teacher Education, 25,* 12–23.

LaBerge, D., & Samuels, S. J. (1974). Toward a theory of automatic information processing in reading. *Cognitive Psychology, 6,* 293–323.

Lampert, M. (1988). What can research on teacher education tell us about improving the quality of mathematics education? *Teaching and Teacher Education: An International Journal of Research and Studies, 3*(4), 157–170.

Land, M. (1979). Low-inference variables and teacher clarity: Effects on student concept learning. *Journal of Educational Psychology, 71*, 795–799.

Lapointe, A. E., Mead, N. A., & Askew, J. M. (1992). *Learning mathematics: An international assessment of educational progress*. Princeton, NJ: ETS.

Lapointe, A. E., Mead, N. A., & Phillips, G. W. (1989). *A world of differences: An international assessment of mathematics and science*. Princeton, NJ: ETS.

Leung, F. K. S. (2005). Some characteristics of East Asian mathematics classrooms based on data from the TIMSS 1999 Video Study. *Educational Studies in Mathematics, 60*(2), 199–215.

Luyten, H. (1998). School effectiveness and student achievement, consistent across subjects? Evidence from Dutch elementary and secondary education. *Educational Research and Evaluation, 4*(4), 281–306.

Luyten, H. (2003). The size of school effects compared to teacher effects: An overview of the research literature. *School Effectiveness and School Improvement, 14*(1), 31–51. doi:10.1076/sesi.14.1.31.13865

Luyten, H., & de Jong, R. (1998). Parallel classes: Differences and similarities: Teacher effects and school effects in secondary schools. *School Effectiveness and School Improvement, 9*(4), 437–473. doi:10.1080/0924345980090405

Ma, L. (1999). *Knowing and teaching elementary mathematics: Teachers' understanding of fundamental mathematics in China and the United States*. Mahwah, NJ: Lawrence Erlbaum.

MacNamara, D. R., & Waugh, D. G. (1993). Classroom organisation: A discussion of grouping strategies in the light of the 'Three Wise Men's' report. *School Organisation, 13*(1), 41–50.

Marton, F., & Booth, S. (1997). *Learning and awareness*. Mahwah, NJ: Lawrence Erlbaum Associates, Inc., Publishers.

McCaleb, J., & White, J. (1980). Critical dimensions in evaluating teacher clarity. *Journal of Classroom Interaction, 15*(2), 27–30.

Meijnen, G. W., Lagerweij, N. W., & Jong, P. F. (2003). Instruction characteristics and cognitive achievement of young children in elementary schools. *School Effectiveness and School Improvement, 14*(2), 159–187. doi:10.1076/sesi.14.2.159.14224

Miller, G. A. (1956). The magic number seven, plus or minus two: Some limits on our capacity for processing information. *Psychological Review, 63*(2), 81–97.

Mortimore, P. (1991). School effectiveness research: Which way at the crossroads? *School Effectiveness and School Improvement, 2*(3), 213–229.

Mortimore, P., Sammons, P., Stoll, L., Lewis, D., & Ecob, R. (1988). *School matters*. Wells: Open books.

Muijs, D., & Dunne, M. (2010). Setting by ability – or is it? A quantitative study of determinants of set placement in English secondary schools. *Educational Research, 52*(4), 391–407.

Muijs, D., Kyriakides, L., van der Werf, G., Creemers, B. P. M., Timperley, H., & Earl, L. (2014). State of the art – teacher effectiveness and professional learning. *School Effectiveness and School Improvement, 25*(2), 231–256. doi:10.1080/09243453.2014.885451

Muijs, D., & Reynolds, D. (2000). School effectiveness and teacher effectiveness in mathematics: Some preliminary findings from the evaluation of the Mathematics Enhancement Programme (Primary). *School Effectiveness and School Improvement, 11*(3), 273–303. Retrieved from www.tandfonline.com/doi/abs/10.1076/0924-3453%28200009%2911%3A3%3B1-G%3BFT273

Muijs, D., & Reynolds, D. (2003). Student background and teacher effects on achievement and attainment in mathematics: A longitudinal study. *Educational Research and Evaluation: An International Journal on Theory and Practice, 9*(3), 289–314.

Muijs, D., & Reynolds, D. (2011). *Effective teaching: Evidence and practice*. (3rd ed.). London: SAGE.

Mullis, I.V. S., Martin, M. O., Beaton, A. E., Gonzalez, E. J., Kelly, D. L., & Smith, T. A. (1997). *Science achievement in the primary school years: IEA's Third International Mathematics and Science Study (TIMSS)*. Chestnut Hill, MA: Center for the Study of Testing, Evaluation and Educational Policy, Boston College.

Mullis, I.V. S., Martin, M. O., & Foy, P. (2008). *TIMSS 2007 international mathematics report: Findings from IEA's Trends in International Mathematics and Science Study at the fourth and eighth grades*. Chestnut Hill, MA: TIMSS & PIRLS International Study Center, Boston College.

Mullis, I.V. S., Martin, M. O., Foy, P., & Arora, A. (2012). *TIMSS 2011 international results in mathematics*. Chestnut Hill, MA: TIMSS & PIRLS International Study Center, Boston College.

Mullis, I.V. S., Martin, M. O., Gonzalez, E. J., & Chrostowski, S. J. (2004). *TIMSS 2003 international mathematics report: Findings from IEA's Trends in International Mathematics and Science Study at the fourth and eighth grades*. Chestnut Hill, MA: TIMSS & PIRLS International Study Center, Boston College.

Mullis, I.V. S., Martin, M. O., Gonzalez, E. J., Garden, R. A., O'Connor, K. M., Chrostowski, S. J., & Smith, T. A. (2000). *TIMSS 1999: International mathematics reports: Finding from IEA's Repeat of Third International Mathematics and Science Study at the eighth grade*. Chestnut Hill, MA: The International Study Center, Boston College.

Mullis, I.V. S., Martin, M. O., Smith, T. A., Garden, R. A., Gregory, K. D., Gonzalez, E. J., . . . O'Connor, K. M. (2003). *TIMSS assessment frameworks and specifications 2003*. (2nd ed.). Chestnut Hill, MA: Boston College.

Munro, J. (1999). Learning more about learning improves teacher effectiveness. *School Effectiveness and School Improvement*, *10*(2), 151–171.

Nanjing Education Bureau. (2011). *Education development in Nanjing in 2011*. Retrieved from www.njedu.gov.cn/default.php?mod=article&do=detail&tid=669644

National Bureau of Statistics of China. (2011). *2010 census of China*. Retrieved from www.stats.gov.cn/tjgb/rkpcgb/qgrkpcgb/t20110428_402722232.htm

National Bureau of Statistics of China. (2013). *Annual data of major cities in China*. Retrieved from http://data.stats.gov.cn/workspace/index?m=csnd

Nuffield Foundation. (1994). *History of mathematics*. Singapore: Longman.

OECD. (2001). *Knowledge and skills for life: First results from the OECD Programme for International Student Assessment (PISA) 2000*. Paris: OECD Publishing.

OECD. (2009a). *Creating effective teaching and learning environments: First results from TALIS*. Paris: OECD Publishing.

OECD. (2009b). *Teaching and Learning International Survey (TALIS): Teacher questionnaire*. Retrieved 7 November 2012 from www.oecd.org/edu/school/43081350.pdf

OECD. (2010). *PISA 2009 results: What students know and can do – student performance in reading, mathematics and science*. (Vol. I). Paris: OECD Publishing.

OECD. (2013). *PISA 2012 results: What students know and can do – student performance in mathematics, reading and science*. (Vol. I). Paris: OECD Publishing.

OECD. (2014). *TALIS 2013 results: An International perspective on teaching and learning*. doi:http://dx.doi.org/10.1787/9789264196261-en

Office for National Statistics. (2013a). *2011 UK censuses*. Retrieved 20 August 2013 from www.ons.gov.uk/ons/guide-method/census/2011/uk-census/index.html

Office for National Statistics. (2013b). *The countries of the UK*. Retrieved 20 August 2013 from http://web.archive.org/web/20081220225201/www.statistics.gov.uk/geography/uk_countries.asp

Opdenakker, M-C., Maulana, R., & den Brok, P. (2012). Teacher – student interpersonal relationships and academic motivation within one school year: Developmental changes and linkage. *School Effectiveness and School Improvement, 23*(1), 95–119.

Oxford Dictionaries. (n.d.). *The definition of pedagogy in English*. Retrieved from www.oxforddictionaries.com/definition/english/pedagogy

Palardy, G. J., & Rumberger, R. W. (2008). Teacher effectiveness in first grade: The importance of background qualifications, attitudes, and instructional practices for student learning. *Educational Evaluation and Policy Analysis, 30*(2), 111–140.

Park, K., & Leung, F. K. S. (2006). Mathematics lessons in Korea: Teaching with systematic variation. In D. Clarke, C. Keitel, & Y. Shimizu (Eds.), *Mathematics classrooms in twelve countries: The insider's perspective* (pp. 247–261). Rotterdam, The Netherland: Sense Publishers.

Peng, W. J., McNess, E., Thomas, S., Wu, X. R., Zhang, C., Li, J. Z., & Tian, H. S. (2014). Emerging perceptions of teacher quality and teacher development in China. *International Journal of Educational Development, 34*, 77–89. doi:http://dx.doi.org/10.1016/j.ijedudev.2013.04.005

Piaget, J. (1950). *The psychology of intelligence*. London: Routledge & Kegan Paul.

Piaget, J. (1962). *Play, dreams and imitation in childhood*. London: Routledge and Kegan Paul.

Piaget, J., & Inhelder, B. (1969). *The psychology of the child*. London: Routledge & Kegan Paul.

Pollard, A., Broadfoot, P., Croll, P., Osborn, M., & Abbott, D. (1994). *Changing English primary schools? The impact of the Education Reform Act at Key Stage 1*. London: Cassell.

Potari, D., Sakonidis, H., Chatzigoula, R., & Manaridis, A. (2010). Teachers' and researchers' collaboration in analysing mathematics teaching: A context for professional reflection and development. *Journal of Mathematics Teacher Education, 13*(6), 473–485. doi:10.1007/s10857-010-9161-9

PRC Gov. (2013). *China profile*. Retrieved 10 September 2013 from www.gov.cn/test/2005-08/11/content_27116.htm

Reid, E. R. (1978–1982). *The reader newsletter*. Salt Lake City, UT: Exemplary Center for Reading Instruction.

Reynolds, D. (1976). The delinquent school. In M. Hammersley & P. Woods (Eds.), *The process of schooling* (pp. 217–219). London: Routledge & Kegan Paul.

Reynolds, D. (Ed.). (1985). *Studying school effectiveness*. London: Falmer.

Reynolds, D. (1995). The effective school: An inaugural lecture. *Evaluation and Research in Education, 9*(2), 57–73.

Reynolds, D. (2010). *Failure free education? The past, present and future of school effectiveness and school improvement*. London: Routledge.

Reynolds, D., Bollen, R., Creemers, B. P. M., Hopkins, D., Stoll, L., & Lagerweij, N. (1996). *Making good schools: Linking school effectiveness and school improvement*. London: Routledge.

Reynolds, D., Creemers, B. P. M., Stringfield, S., Teddlie, C., & Schaffer, G. (Eds.). (2002a). *World class schools: International perspectives on school effectiveness*. London: Routledge.

Reynolds, D., & Farrell, S. (1996). *Worlds apart? A review of international surveys of educational achievement involving England*. London: H.M.S.O.

Reynolds, D., Sammons, P., De Fraine, B., Townsend, T., & Van Damme, J. (2011). *Educational effectiveness research: A state of the art review*. Paper presented at The Annual Meeting of the International Congress for School Effectiveness and Improvement, Cyprus.

Reynolds, D., Stringfield, S., Teddlie, C., & Creemers, B. P. M. (2002b). The intellectual and policy context. In D. Reynolds, B. P. M. Creemers, S. Stringfield, C. Teddlie, & G. Schaffer (Eds.), *World class schools: International perspective on school effectiveness* (pp. 3–14). London: Routledge Falmer.

Reynolds, D., & Teddlie, C. (2000). The processes of school effectiveness. In C. Teddlie & D. Reynolds (Eds.), *The international handbook of school effectiveness research* (pp. 134–159). London: Falmer Press.

Robitaille, D. F., & Garden, R. A. (Eds.). (1989). *The IEA study of mathematics II: Contexts and outcomes of school mathematics.* Oxford: Pergamon Press.

Rosenshine, B. (1971). *Teaching behaviours and student achievement.* London: National Foundation and Educational Research.

Rosenshine, B., & Furst, N. (1973). The use of direct observation to study teaching. In R. M. W. Travers (Ed.), *Second handbook of research on teaching* (pp. 122–183). Chicago, IL: Rand McNally.

Rosenshine, B., & Stevens, R. (1986). Teaching functions. In M. C. Wittrock (Ed.), *Handbook of research on teaching* (3rd ed., pp. 376–391). New York: Palgrave Macmillan.

Rossella, S., & Alessandra, B. (2005). Mathematics teaching in Italy: A cross-cultural video analysis. *Mathematical Thinking & Learning,* 7(4), 291–312.

Rowland, T. (2010). Foundation knowledge for teaching: Contrasting elementary and secondary mathematics. In V. Durand-Guerrier, S. Soury-Lavergne, & F. Arzarello (Eds.), *Proceedings of the sixth congress of the European society for research in mathematics education* (pp. 1841–1850). Lyon: CERME.

Ruddock, G., Sainsbury, M., Clausen-May, T., Vppula, H., Mason, K., & Patterson, E. W. (2008). *Comparison of the core primary curriculum in England to those of other high performing countries.* London: DCSF.

Rutter, M., Maughan, B., Mortimore, P., & Ouston, J. W., Smith, A. (1979). *Fifteen thousand hours: Secondary schools and their effects on children.* Boston, MA: Harvard University Press.

Sammons, P. (1999). *School effectiveness: Coming of age in the 21st century.* Liss: Swets & Zeitlinger.

Sammons, P., Thomas, S., & Mortimore, P. (1997). *Forging links: Effective departments and effective schools.* London: Paul Chapman.

Scheerens, J., & Bosker, R. (1997). *The foundations of school effectiveness.* Oxford: Pergamon Press.

Scheerens, J., Vermeulen, C. J. A. J., & Pelgrum, W. J. (1989). Generalizability of instructional and school effectiveness indicators across nations. *International Journal of Educational Research, 13*(7), 789–799.

Schmidt, W. H., Jorde, D., Cogan, L. S., Barrier, E., Gonzalo, I., Moser, U., . . . Wolfe, R. G. (1996). *Characterizing pedagogical flow: An investigation of mathematics and science teaching in six countries.* Dordrecht, The Netherlands: Kluwer Academic Publishers.

Schmidt, W. H., McKnight, C. C., Houang, R. T., Wang, H. C., Wiley, D. E., Cogen, L. S., & Wolfe, R. G. (2001). *Why schools matter: A cross-national comparison of curriculum and learning.* San Francisco, CA: Jossey-Bass.

Schön, D. A. (1983). *The reflective practitioner: How professionals think in action.* New York: Basic Books.

Schön, D. A. (1987). *Educating the reflective practitioner toward a new design for teaching and learning in the professions.* San Francisco, CA: Jossey-Bass.

Sekiguchi, Y. (2006). Coherence of mathematics lessons in Japanese eighth-grade classrooms. In J. Novotná, H. Moraová, M. Krátká, & N. Stehlíková (Eds.), *Proceedings of 30th conference of the international group for the psychology of mathematics education* (Vol. 5, pp. 81–88). Prague, Czech Republic: PME.

Shimizu, Y. (2007). Explicit linking in the sequence of consecutive lessons in mathematics classroom in Japan In J.-H. Woo, H.-C. Lew, K.-S. Park & D.-Y. Seo (Eds.), *Proceedings of the 31st conference of the international group for the psychology of mathematics education* (Vol. 4, pp. 177–184). Seoul, South Korea: PME.

Shulman, L. (1986). Those who understand: Knowledge growth in teaching. *Educational Researcher, 15(2)*, 4–14.

Shulman, L. (1987). Knowledge and teaching: Foundations of the new reform. *Harvard Educational Review, 57(1)*, 1–22.

Smith, D. J., & Tomlinson, S. (1989). *The school effect: A study of multi-racial comprehensives*. London: Policy Studies Institute.

Smith, L., & Cotton, M. (1980). Effect of lesson vagueness and discontinuity on student achievement and attitudes. *Journal of Educational Psychology*, *72*, 670–675.

Smith, L., & Land, M. (1981). Low-inference verbal behaviours related to teacher clarity. *Journal of Classroom Interaction*, *17*(1), 37–42.

Soar, R. S. (1966). *An integrative approach to classroom learning*. Philadelphia, PA: Temple University.

Soar, R. S. (1973). *Follow through classroom process measurement and pupil growth*. Gainesville, FL: University of Florida.

Soar, R. S. (1977). An integration of findings from four studies of teacher effectiveness. In G. Borich & K.S. Fenton (Eds.), *The appraisal of teaching: Concepts and process* (pp. 146–161). New York: Addison-Wesley.

Soar, R. S., & Soar, R. M. (1972). An empirical analysis of selected Follow Through programmes: An example of a process approach to evaluation. In I. Gordon (Ed.), *Early childhood education* (pp. 229–259). Chicago, IL: National Society for the Study of Education.

Soar, R. S., & Soar, R. M. (1973). *Classroom behaviour, pupil characteristics and pupil growth for the school year and the summer*. Gainesville, FL: University of Florida, Institute for Development of Human Resources.

Soar, R. S., & Soar, R. M. (1978). *Setting variables, classroom interaction, and multiple pupil outcomes*. Washington, DC: National Institute of Education.

Southampton City Council. (2013). *Southampton city statistics and research*. Retrieved 10 August 2013 from www.southampton.gov.uk/living/statsresearch/

Stallings, J., Cory, R., Fairweather, J., & Needles, M. (1977). *Early childhood education classroom evaluation*. Menlo Park, CA: SRI International.

Stallings, J., Cory, R., Fairweather, J., & Needles, M. (1978). *A study of basic reading skills taught in secondary schools*. Menlo Park, CA: SRI International.

Stallings, J., & Kaskowitz, D. (1974). *Follow through classroom observation evaluation 1972–1973*. Stanford, CA: Stanford Research Institute.

Stevenson, H. W., & Stigler, J. W. (1992). *The learning gap: Why our schools are failing, and what we can learn from Japanese and Chinese education*. New York: Summit Books.

Stigler, J. W., Gonzales, P., Kawanaka, T., Knoll, S., & Serrano, A. (1999). *The TIMSS videotape classroom study: Methods and findings from an exploratory research project on eighth-grade mathematics instruction in Germany, Japan, and the United States*. Washington, DC: National Centre for Education Statistics.

Stigler, J. W., & Hiebert, J. (1999). *The teaching gap: Best ideas from the world's teachers for improving education in the classroom*. New York: The Free Press.

Stigler, J. W., & Perry, M. (1988). Mathematics learning in Japanese, Chinese, and American classrooms. *New Directions for Child and Adolescent Development, 1988*(41), 27–54.

Stockero, S. L. (2008). Using a video-based curriculum to develop a reflective stance in prospective mathematics teachers. *Journal of Mathematics Teacher Education, 11*(5), 373–394. doi:10.1007/s10857-008-9079-7

Teddlie, C., Creemers, B. P. M., Kyriakides, L., Muijs, D., & Fen, Y. (2006). The international system for teacher observation and feedback: Evolution of an international study of teacher effectiveness constructs. *Educational Research and Evaluation, 12*(6), 561–582.

Teddlie, C., & Liu, S. (2008). Examining teacher effectiveness within differentially effective primary schools in the People's Republic of China. *School Effectiveness and School Improvement, 19*(4), 387–407. doi:10.1080/09243450802535182
Teddlie, C., & Reynolds, D. (2000). *The international handbook of school effectiveness research*. London: Falmer press.
Teddlie, C., & Stringfield, S. (1993). *Schools make a difference: Lessons learned from a ten year study of school effects*. New York: Teachers College Press.
Teodorović, J. (2011). Classroom and school factors related to student achievement: What works for students? *School Effectiveness and School Improvement, 22*(2), 215–236. doi:10.1080/09243453.2011.575650
Tikunoff, W., Berliner, D., & Rist, R. (1975). *An ethnographic study of the forty classrooms of the beginning teacher evaluation study known sample*. San Francisco, CA: Far West Laboratory.
Tizard, B., Blatchford, P., Burke, J., Farquhar, C., & Plewis, I. (1988). *Young children at school in the inner city*. Hove: Lawrence Erlbaum Associate Ltd.
Tobin, J. J., Hsueh, Y., & Karasawa, M. (2009). *Preschool in three cultures revisited: China, Japan, and the United States*. Chicago, IL: The University of Chicago Press.
Tobin, J. J., Wu, D. Y. H., & Davidson, D. H. (1989). *Preschools in three cultures: Japan, China, and the United States*. New Haven, CT: Yale University Press.
The World Bank. (2013). *GDP per capita*. Retrieved 20 August 2013 from http://data.worldbank.org/indicator/NY.GDP.PCAP.CD/countries?order=wbapi_data_value_2012 wbapi_data_value wbapi_data_value-last&sort=desc&display=default
Van Damme, J., Opdenakker, M., Van Landeghem, G., De Fraine, B., Pustjens, H., & Van de gaer, E. (2006). *Educational effectiveness: An introduction to international and Flemish research on school, teachers and classes*. Leuven: Acco.
Van de gaer, E., De Fraine, B., Van Damme, J., De Munter, A., & Onghena, P. (2009). School effects on the development of motivation toward learning tasks and the development of academic self-concept in secondary school. *School Effectiveness and School Improvement, 20*(2), 235–253.
Van der Werf, G., Opdenakker, M-C., & Kuyper, H. (2008). Testing a dynamic model of student and school effectiveness with a multivariate multilevel latent growth curve approach. *School Effectiveness and School Improvement, 19*(4), 447–462.
Van Landeghem, G., Van Damme, J., Opdenakker, M-C., De Fraine, B., & Onghena, P. (2002). The effect of schools and classes on noncognitive outcomes. *School Effectiveness and School Improvement, 13*(4), 429–451.
Verachtert, P., Van Damme, J., Onghena, P., & Ghesquiere, P. (2009). A seasonal perspective on school effectiveness: Evidence from a Flemish longitudinal study in kindergarten and first grade. *School Effectiveness and School Improvement, 20*(2), 215–233.
Walker, D., & Schaffarzick, J. (1974). Comparing curricula. *Review of Educational Research, 44*, 83–111.
Walshaw, M. (2010). Mathematics pedagogical change: Rethinking identity and reflective practice. *Journal of Mathematics Teacher Education, 13*(6), 487–497. doi:10.1007/s10857-010-9163-7
Weber, G. (1971). *Inner city children can be taught to read: Four successful schools*. Washington, DC: Council for Bacis Education.
Wikipedia. (2013). *List of countries by GDP (nominal) per capita*. Retrieved 10 August 2013 from http://en.wikipedia.org/wiki/List_of_countries_by_GDP_(nominal)_per_capita
Willms, J. D. (2000). Monitoring school performance for standards-based reform. *Evaluation and Research in Education, 14*(3–4), 237–253.

Index

For Product Safety Concerns and Information please contact our EU representative GPSR@taylorandfrancis.com
Taylor & Francis Verlag GmbH, Kaufingerstraße 24, 80331 München, Germany

www.ingramcontent.com/pod-product-compliance
Ingram Content Group UK Ltd.
Pitfield, Milton Keynes, MK11 3LW, UK
UKHW022125190926
475702UK00013B/1660
9780367178635